Praise for *Heaven and Earth*

'Big in theme, languid in pace and exquisite in execution . . . The plot is deftly handled, moving from a secretive steamy teenage romance in Speziale to a cave in Iceland – taking in fringe eco-activism and a doomed attempt to conceive a child along the way . . . The dreamy lyricism of the prose ("the foam-slick rocks, the silent sea, and, all around, the mercilessly bright night of the South") . . . Giordano's novel is a devastating marvel'   *Sunday Telegraph*

'*Heaven and Earth* is rooted so deep in idyllic Puglia that you can almost feel the red soil under your sandals'   *Daily Mail*

'A highly enjoyable novel, convincingly and smoothly translated by Anne Milano Appel . . . Giordano is especially good on the textures, smells, heat and colours of the Italian south, where almost the whole novel is set, the herbs that scent the air, the rocky terrain on which little grows. These stay long in the mind, as does the way he writes about the obsessiveness of love, the way it dominates and distorts and the self-delusions and fantasies it gives rise to. Puglia's scorched earth and, later in the novel, the craters and caverns of Iceland become metaphors for a plot that is both touching and sad, violent and uncomfortable'   *TLS*

'Giordano is a fluid, expansive writer. The chapters flow effortlessly back and forth in time, pulling us deeper into the story of Teresa and Bern's great love. The landscape shimmers with their longing'
                                                                    New York Times

'It's been too long since Italian author Paolo Giordano (who happens to have a PhD in particle physics) wrote a novel . . . [H...]
in Puglia and focuses on four [...]
sprawls and stuns'

'This is at once a lush picture of growing up in the Italian countryside and a deeply affecting story of friendships under the strain of time and tragedy. Giordano's best book yet'

Dave Eggers, bestselling author of *The Circle*

'An intense novel about passions and reasons, unbreakable bounds and reckless excursions. Giordano is a master storyteller'

Yiyun Li, author of *Where Reasons End*

'Magnificent, heart-wrenching, and utterly compelling. *Heaven and Earth* is the perfect novel. And I'm not saying this lightly. Perfect. Perfect. Perfect'

Andrea Wulf, Costa-winning author of *The Invention of Nature*

'*Heaven and Earth* is not just a magnificent novel – it's an act of faith in literature and in the rousing power of storytelling, an ode to the unknowable mystery that is the human heart. Novels like this are a rare find: you won't be able to forget it'

Elena Varvello, bestselling author of *Can You Hear Me?*

'A novel as ferocious as youth and as pure as a utopia'

Paolo Cognetti, award-winning author of *The Eight Mountains*

Praise for Paolo Giordano

'Mesmerizing . . . Giordano works with piercing subtlety. An exquisite rendering of what one might call feelings at the subatomic level'

New York Times on *The Solitude of Prime Numbers*

Heaven
and Earth

Paolo Giordano is a physicist and a bestselling writer. His first novel, *The Solitude of Prime Numbers*, was translated into more than 40 languages worldwide and won the Premio Strega (the Italian Booker). He is also the author of the pamphlet *How Contagion Works*, a rallying cry inspired by the recent Covid-19 pandemic, and the creator of the HBO series *We Are Who We Are*, co-written and directed by Luca Guadagnino. He holds a master's degree and a PhD in theoretical physics.

ALSO BY PAOLO GIORDANO

The Solitude of Prime Numbers
The Human Body
Like Family

How Contagion Works:
Science, Awareness and Community in Times of Global Crises

HEAVEN
AND EARTH

PAOLO GIORDANO

English Translation by Anne Milano Appel

WEIDENFELD & NICOLSON

First published in Great Britain in 2020 by Weidenfeld & Nicolson
This paperback edition published in Great Britain in 2021 by Weidenfeld & Nicolson
an imprint of The Orion Publishing Group Ltd
Carmelite House, 50 Victoria Embankment
London EC4Y 0DZ
An Hachette UK Company

First published in Italy as *Divorare il Cielo* by Giulio Einaudi editore, Turin, Italy in 2018.

1 3 5 7 9 10 8 6 4 2

Copyright © Paolo Giordano 2018
English translation © Anne Milano Appel 2020

A CIP catalogue record for this book is
available from the British Library.

ISBN (Paperback) 978 1 4746 1216 6
ISBN (eBook) 978 1 4746 1217 3

Typeset by Input Data Services Ltd, Somerset

Printed in Great Britain by Clays Ltd, Elcograf S.p.A.

www.weidenfeldandnicolson.co.uk
www.orionbooks.co.uk

To Rosaria and Mimino,

to Angelo and Margherita.

To their songs.

CONTENTS

PART ONE

THE GREAT
EGOISTS

1.

I saw them swimming in the pool, at night. There were three of them and they were very young, barely children, as I too was then—just a girl.

In Speziale, my sleep was continually interrupted by new sounds: the swooshing of the irrigation system, the feral cats that tussled in the grass, a bird that made the same sound over and over again. In the first summers spent at my grandmother's, it seemed I almost never slept. From the bed where I lay, I stared at the objects in the room as they receded and drew near, as if the whole house were breathing. That night I heard noises in the yard, but I didn't move right away; sometimes the watchman came up to the house to leave a note stuck to the door. But then there were whispers and muffled laughter and I decided to get up.

Making sure my feet avoided the mosquito coil that glowed from the floor, I went to the window and looked down: I was too late to see the boys undressing, but in time to catch the last of them slipping into the black water.

I could make out their heads, two that were darker and one that looked like silver. Apart from that, seen from where I stood, they were almost identical, moving their arms in circles to keep afloat.

There was a kind of tranquility in the air, after the tramontana had subsided. One of the boys started playing dead man in the middle of the pool. I felt my throat burn at the sudden sight of his nakedness, even though he was only a shadow, more my imagination than anything else. He arched his back and somersaulted into a dive. When he reemerged and gave a shout, his silver-haired friend slapped his face to shut him up.

"You hurt me, you idiot!" the one who had somersaulted said loudly.

The other boy shoved him under the water, then the third one also jumped on him. I was afraid they were beating one another up, that someone might drown, but instead they broke apart, laughing. They sat on the edge of the pool over by the shallow end, their wet backs turned to me. The boy in the center, the taller one, reached out and draped his arms around the others' shoulders. They talked quietly, but I was able to catch a few words here and there.

For a moment I thought about going down and sinking into the steamy night with them. The seclusion I felt at Speziale left me hungry for any human contact, but at fourteen I didn't have the nerve for certain things. I suspected that they were the boys from the neighboring property, even though I had only ever seen them from a distance. My grandmother called them "the kids from the masseria."

Then the creaking of bed springs. A cough. My father's rubber flip-flops slapping on the floor. Before I could call out to the boys to run, he was rushing down the stairs, calling the caretaker. The light went on in the làmia, the caretaker's lodge, and Cosimo came out at the same instant my father appeared in the yard, both in their boxers.

The boys had jumped out of the pool and were grabbing their scattered clothes. Leaving some behind on the ground, they started running into the darkness. Cosimo started after them, shouting, I'll kill you, you little bastards, I'll beat your brains out, and after a moment's hesitation my father followed him. I saw him pick up a rock.

From the darkness came a cry, then the smack of bodies against the fence, a voice barking no, get down from there. My heart was pounding rapidly, as if I were the one running away, the one being chased.

It was a while before they came back. My father was holding his left wrist, he had a mark on his hand. Cosimo examined it closely, then pushed him into the lodge. Before he too disappeared into the house, he stared a moment at the darkness that had swallowed the invaders.

THE NEXT DAY, at lunch, my father's hand was bandaged. He said he'd stumbled while trying to put back a magpie nest. In Speziale he turned into a different person: in just a few days his skin became very dark and even his voice changed with the dialect. I felt as if I didn't know him at all. Sometimes I wondered who he really was: the engineer who in Turin always wore a suit and tie, or the man with the unkempt beard who went around my grandmother's house half naked. In any case, it was clear that my mother had chosen to marry only one of the two, and that she wanted nothing to do with the other one. She had not set foot in Puglia in years. At the beginning of August, when my father and I left to face the interminable car journey to the South, she didn't even come out of her room to say goodbye.

We ate in silence, until we heard Cosimo's voice calling from the yard.

On the threshold, in front of the caretaker, were the three boys from the night before. At first I recognized only the tall one, because of his thin neck and the oblong shape of the head. But then my attention was drawn to the other two. One had very fair skin, with hair and eyebrows as white as cotton; the other one was dark-haired, tanned, and his arms were scored with scratches.

"So," my father said, "have you come to get your clothes?"

The tall one replied flatly: "We came to apologize for entering your property last night and for using the pool. Our parents send you these." He held out a bag and my father took it.

"What's your name?" he said. In spite of himself, he had softened slightly.

"Nicola."

"And them?"

"He's Tommaso," he said, pointing to the albino one. "And he's Bern."

They looked uncomfortable in their T-shirts, as if someone had made them wear them. I exchanged a long look with Bern. He had dark, close-set eyes.

My father jiggled the bag a little and the jars inside clinked.

"It wasn't necessary for you to sneak in," he said. "If you wanted to use the pool, all you had to do was ask."

Nicola and Tommaso lowered their gazes, while Bern continued to stare at me. The white stone of the patio behind them was dazzling.

"If something had happened to one of you . . ." My father broke off, more and more embarrassed. "Cosimo, did we offer these boys some lemonade?"

The caretaker rolled his eyes, as if asking him if he had lost his mind.

"That's okay, thank you," Nicola said politely.

"If your parents will allow it, this afternoon you can come and have a swim." My father glanced at me, perhaps to ask for my consent.

At that point Bern spoke up: "Last night you hit Tommaso from behind with a rock. We committed an offense by entering your property, but you committed a more serious one by injuring a minor. If we wanted to, we could report you."

Nicola elbowed him in the chest.

"I didn't do any such thing," my father replied. "I don't know what you're talking about."

The image of him bending down to pick up the rock came back to me and again I heard the sounds coming from the darkness, that cry that I had not been able to make out.

"Tommi, show Signor Gasparro the bruise, please."

Tommaso drew back, but when Bern reached for the edge of his T-shirt, he did not resist. Gently, Bern rolled up the fabric, uncovering his back: it was even paler than his arms. The pallor made the bluish contusion, the size of a glass, stand out.

"See?"

Bern pressed his index finger on the bruise, and Tommaso squirmed free.

My father seemed dazed. Cosimo intervened in his place; he issued an order to the boys in dialect and they calmly said goodbye with a bow.

When they were already in full sunlight, Bern turned around to cast a stern look at our house. "I hope your hand heals quickly," he said.

THAT AFTERNOON a thunderstorm struck. In a few minutes the sky turned purple and black, colors I'd never seen before.

The rain lasted for almost a week; the clouds came from the sea, out of the blue. A lightning strike splintered a branch of the eucalyptus tree and another incinerated the pump that drew water from the well. My father was furious and took it out on Cosimo.

My grandmother, on the sofa, read her paperback thrillers. Just to pass the time, I asked her to recommend one to me. She told me to pick one at random from the bookcase, they were all good. I chose *Deadly Safari*, but the story was boring.

After staring blankly into space for a while, I asked her what she knew about the boys from the masseria.

"They come and go," she said. "They're never the same ones for too long."

"And what do they do?"

"They wait for their parents to take them back, I imagine. Or someone else who will."

She put down the book. "Meanwhile, they pray. They're part of some kind of . . . sect."

When the bad weather ended, there was an invasion of frogs. At night they leaped into the pool, and no matter how much chlorine we added, there was no way to keep them out. We found them trapped in the filters or crushed by the wheels of the pool cleaner. Those who survived swam along peacefully, some in pairs, one clinging to the other's back.

One morning I went down to the patio for breakfast, still in my pajama shorts and tank top, and I saw Bern. From the edge of the pool he was pursuing the frogs with a net. When he caught one, he let the water drain out, then overturned the creature into a bucket.

For a while I wasn't sure whether to let him see me or go back in and get dressed, but in the end I went over to him and asked him if my father paid him to do that job.

"Cesare doesn't like us handling money," he said, barely turning around. After a pause he added: "'Then one of the Twelve went to the chief priests and said: "What are you willing to give me if I hand him over to you?" And they paid him thirty pieces of silver.'"

It seemed like a nonsensical answer, but I didn't feel like having him explain it to me. I looked into the bucket: the piled-up frogs leaped upward, but the plastic walls were too high.

"What are you going to do with them?"

"Let them go."

"If you let them go, they'll come back tonight. Cosimo kills them with baking soda." Quick as a flash, Bern looked up. "I'll take them far enough away, you'll see."

I shrugged. "I don't know why you're doing this lousy job if you don't even get paid for it."

"It's my punishment, for using your pool without permission."

"You already apologized."

"Cesare said we had to make up for it. But until today I haven't had the chance because of the rain."

In the water, the frogs were hurriedly fleeing. He pursued them patiently with the net.

"Who is Cesare?"

"Nicola's father."

"Is he your father, too?"

Bern shook his head. "He's my uncle."

"What about Tommaso? Is he your brother?"

Again he shook his head no. When they had introduced themselves at our door, Nicola had said "our parents." But Bern probably wouldn't make it easy for me to understand and I didn't want to give him the satisfaction.

"How's his bruise?" I asked.

"It hurts to raise his arm. At night Floriana makes apple vinegar compresses for him."

"Anyway, I think you were wrong, it wasn't my father who threw the rock. It must have been Cosimo."

Bern didn't seem to be listening to me, he was wholly absorbed in fishing out the frogs. He was wearing pants that at one time must have been blue, and he was barefoot. Then, point-blank, he said: "You really have some nerve."

"What?"

"Accusing Signor Cosimo to excuse your father. I don't think you pay him enough for that."

Another frog plopped into the bucket. There must have been about twenty or so, they swelled up and deflated.

I wanted to cover up my earlier lie, so I asked: "How come your friends didn't come?"

"It was my idea to use the pool."

I felt my hair, it was broiling. I could have leaned over, dipped my hand in the water, and wet my head, but there were still frogs in the pool.

Bern scooped one up and held the net out to me. "Want to touch it?"

"No way!"

"I figured," he said with a nasty smile. Then he added: "Tommaso went to see his father in prison today."

He waited for me to react, but I remained silent.

"He killed his wife with a wooden clog. Afterward he tried to hang himself from a tree, but the police caught him in time."

The frogs thumped frantically against the bucket. All those slimy bodies crammed together.

"You're making that up, right?"

Bern stopped, the net suspended. "Of course not."

Finally he caught the last frog, the one that had been giving him a hard time. He got down on his knees so he wouldn't have to lift the net up too far.

"And your parents?" I asked.

The frog escaped with a leap and plunged into the deepest part of the pool.

"Damn! See what you did? You're an *arruffona!*"

I lost my patience. "And what is an *arruffona*, huh? That's not even a real word! Look, I wasn't the one who hurt your brother, or your friend, whatever the hell he is!"

I was determined to walk off immediately, but for the first time Bern looked at me seriously. His face expressed a sincere remorse, and at the same time a kind of naiveté. And there were his close-set eyes, again.

"Please accept my apologies," he said.

I was a little uncomfortable, like the week before, when he'd stared at me over my father's shoulders. I leaned over the water to see where the frog had gone to.

"What are those black strands?"

"Eggs. The frogs came here to deposit them."

"That's awful."

But he misunderstood what I meant.

"Yes, it's awful. Not only are you killing the frogs, but also all those eggs. Inside each one there's a living creature."

LATER I LAY DOWN to sunbathe, but it was two o'clock, the worst time of day, and I couldn't stand it for long. I crossed the yard and went past the stones that separated it from the open countryside. Where the boys had climbed over, the fence was bent at the top and buckled in the center. Beyond it were more trees, a little taller than ours. I tried to see the masseria, but it was too far away.

Before leaving, Bern had invited me to the burial of the frogs that he'd fished out dead.

I'd asked Cosimo to inflate the tires of my grandmother's old bicycle, and he had it ready for me in the yard, oiled and gleaming.

"Where are you going?"

"Just for a little ride, around here, along the road." I waited for my father to go out to meet his friends, then I set off.

The entrance to the masseria was on the side opposite ours; to get there you had to go all the way around, unless you decided to climb over and cut through the fields, as the boys had done. Along the stretch of paved road, trucks darted past me. I had propped the Walkman in the basket and had to keep leaning forward because the headphones' cord was too short.

The masseria didn't have a real gate, just an iron bar, which I found open. Weeds grew in the middle of the dirt track. I hopped off my bike and continued on foot. It took me another five minutes to reach the house.

I left the bike on its side and cleared my throat, but nobody appeared. I walked a few steps more to get out of the sun and into the shade of a pergola. The door to the house was wide open, but I didn't feel I should just go in. I leaned over the table instead, intrigued by the plastic tablecloth that depicted a world map. I looked for Turin, but it wasn't there.

I put my headphones back on and walked around the house, peering through the windows, but the contrast between the darkness inside and the light outside was too stark. Then, in the back, I found Bern.

He was sitting on a stool, in a shady corner, bent over the ground. In that position his vertebrae formed a strip of bumps down the middle of his back. Surrounding him were heaps of almonds, an infinity of almonds, so many that I could have lain down on top of them, arms outstretched, and been swallowed up.

He didn't notice me until I was in front of him, and even then he gave no sign of being surprised.

"Here she is, the rock thrower's daughter," he murmured.

A surge of humiliation rose from my stomach. "My name is Teresa." He hadn't asked me the whole time we'd been together that morning. He nodded, but as if the information were of no interest to him.

"What are you doing?" I asked.

"Can't you see?"

He grabbed a handful of four or five almonds, removed the hulls, and dropped them into a separate pile.

"Are you going to husk them all?"

"Yep."

"That's crazy, there must be thousands of them."

"You could help me, instead of standing there, doing nothing."

"Where should I sit?" Bern shrugged. I settled on the ground with my legs crossed.

We hulled almonds for a while. I noticed how many he had already husked; he must have been sitting there for hours.

"You're very slow," he said at one point.

"Well, it's the first time I'm doing it!"

"That doesn't matter, you're slow, period."

"You said we would bury the frogs."

"I said at six."

"I thought it was already six," I lied.

Bern glanced at the sun, uncramped his neck. I reached out listlessly to grab another handful. The trick to removing the hull more quickly was not to worry about the pulp ending up under your fingernails.

"Did you pick all of them yourself?"

"All of them, yes."

"And what are you planning to do with them?"

Bern sighed. "On Sunday my mother is coming. She loves almonds, but they need at least two days to dry in the sun. And then you have to crack the shells, which takes even longer. So I'm late. I have to finish them by tomorrow."

I stopped. I was already tired and the pile hadn't grown any smaller. I fidgeted a little to get Bern's attention, but he didn't take his eyes off the ground.

"Do you like the new song by Roxette?" I asked him.

"Sure."

But I had the impression it wasn't true, that he didn't know the song at all, or Roxette either. After a while he said, "Is that what you were listening to?"

"Do you want to hear it?"

Bern hesitated before dropping the almonds. I gave him the

Walkman. He put the headphones on and started turning the recorder in his hands.

"You have to press play."

He examined it again, one side and the other, then returned it to me with a nervous gesture.

"It doesn't matter."

"Why? I'll show you how . . ."

"It doesn't matter."

We went on working, not looking at each other and not speaking anymore, only the clicks of the hulled almonds, clack, clack, clack, until the other boys came looking for us.

"What is she doing here?" Tommaso asked.

Bern stood up to face him. "I told her to come."

Nicola, more politely, held out his hand and introduced himself, as if I couldn't remember his name. I wondered which one of the three had played dead man in the pool. It was as if what I had seen that night gave me an unfair advantage over all of them.

Then Tommaso said: "He's ready over there, let's go," and he set off without us.

A man was waiting for us in a clearing among the olive trees. "Welcome, my dear," he said to me, opening his arms.

A stole with two gold-embroidered crosses was draped over his shoulders and in one hand he held a small leather-bound book. He had a black beard, and his eyes were a very pale blue, almost transparent. "I am Cesare."

Five small holes had been dug at his feet, and the frogs were already in them. Cesare patiently explained to me what was happening: "Man buries his dead, Teresa, he always has. This is how our civilization began, and this is how souls are assured of their journey to a new haven. Or to Jesus, if their cycle has been completed."

When he said "Jesus," they all made the sign of the cross, twice in a row, kissing their thumbnails at the end.

In the meantime, a woman had approached, holding a guitar by its neck; she stroked my cheek, as if she had known me forever.

"Do you know what the soul is?" Cesare asked.

"I'm not sure."

"Have you ever seen a plant that is about to die? Maybe of thirst?"

In Turin our neighbors' sentry palm had dried up on the balcony; the owners had gone on vacation without thinking about it. I nodded.

"At a certain point," he went on, "the leaves shrivel up, the branches droop, and the plant becomes a pitiable thing. Life has already left it. The same thing happens to our body when the soul leaves it," he said, his head coming a little closer to mine. "But there is something they didn't teach you in catechism class. We do not die, Teresa, because souls migrate. Each of us has many lives behind us and many more ahead of us, as a man, a woman, or an animal. Even these poor frogs. This is why we want to bury them. It doesn't cost us much, does it?"

He stared at me, satisfied, then, without looking away, he said: "Floriana, whenever you're ready."

The woman took up the guitar. Since she did not have a shoulder strap, she had to bend a knee to support it. She started strumming arpeggios and singing a gentle song.

After a few moments, the others joined in. Bern was the only one who kept his eyes closed and his chin slightly raised. I would have liked to hear him singing by himself, at least for a moment.

At a certain point they joined hands. Cesare, who was on my left, gave me his. I didn't know what to do about Floriana, who was playing the guitar. I saw that Tommaso was resting his fingers on her shoulder; not wanting to break the circle, I did that, too, and she smiled at me.

The frogs were rigid, withered, there couldn't really be a soul in those viscous bellies. I wondered whether Cesare believed it was still there or it had already flown away elsewhere. In any case, the bodies were blessed and the boys knelt down to fill the holes back up. "They're part of some kind of sect," my grandmother had said.

Before I left, Cesare invited me to come back: "We have so many things to talk about, Teresa."

ALONG THE DIRT TRACK Bern pushed the bicycle by the handlebars for me. "So did you like it?" he asked me.

I said yes, mainly to be polite. Only later did I realize it was true.

"'Not for your sacrifices do I rebuke you,'" he said, "'your burnt offerings are always before me.'"

"What?"

"'I will not take a bullock from your house, or he-goats from your folds,'" he went on, repeating one of the prayers that Cesare had read shortly before. "'I know every bird in the heights; whatever moves in the wild is mine.' It's my favorite verse, when he says 'whatever moves in the wild is mine.'"

"You know it by heart?"

"I've learned some of the psalms by heart, though not all of them yet," he explained, almost apologetically.

"Why?"

"Because I haven't had time!"

"No, I meant why do you memorize them? What's the purpose?"

"The psalms are the only way to pray, the only way that pleases God."

"Is it Cesare who teaches you these things?"

"He teaches us everything."

"You three don't go to a normal school, do you?"

As he wheeled the bicycle over a stone, the chain wobbled.

"Careful!" I told him. "Cosimo just fixed it."

"Cesare knows a lot more things than what you learn in normal schools, as you call them. He was an explorer when he was young, he lived in Tibet, alone, in a cave. He thinks that at some point he didn't even feel the cold anymore, he could easily survive at four degrees below zero without clothes. And he hardly ate anything. That's where he discovered metempsychosis."

"Discovered what?"

"The transmigration of souls. It's spoken of in many passages of the Gospels, in Matthew, for example. But above all, in John."

"And you really believe it?"

He looked at me sternly. "I bet you haven't read even one page from the Bible."

We had reached the iron bar at the end of the track and he stopped abruptly. Handing the bicycle back to me, he said, "You can come again if you want. After lunch, the others nap, there's only me."

SOMETIMES I WONDER why I went back to the masseria, if it was because I wanted to see Bern again—a curiosity that did not yet have a name—or simply because I was bored in Speziale. In any case, I returned the following afternoon, helped him with the almonds, and together we managed to hull them all.

On my last day in Puglia, I spent the whole morning gathering up my things and packing them into my suitcase. Usually I was thrilled at the idea of leaving, but not that year. After lunch, I took the bike and pedaled over to the masseria.

Bern wasn't there, however. I walked around the house twice, whispering his name. The almonds were still there; without their husks, they looked strangely diminished.

Back under the pergola, I sat down on the two-seater swing-chair and gave it a slight push. Two cats slept, lying on their sides, stunned by the heat. Then I heard my name being called.

"Where are you?" I asked softly.

Bern directed my gaze toward a window on the second floor, he, too, whispering: "Come closer."

"Why don't you come down?"

"I can't get out of bed. My back seized."

I thought about all the hours he had spent hunched over the almonds. "Can I come up there?"

"Better not. You'd wake Cesare."

I felt like a fool, talking to a window.

"I wanted to give you something. I'm leaving tonight."

"Where are you going?"

"Back home. To Turin."

Bern was silent for a few moments. Then he said: "Safe trip, then."

Maybe he would leave during the winter, and I wouldn't see him again. "They come and go," my grandmother had said. A beetle crawled near my foot; I crushed it under the sole of my sandals. Would they bury that, too?

I picked my bicycle up from the ground. I was already straddling it when Bern called me again.

"Now what?"

"You can take some almonds. Bring them with you to Turin."

"Why, didn't your mother want them?"

I wanted to be rude and I probably succeeded. He seemed to think for a moment.

"Take them," he said finally, "as many as you want. Put them in the bike basket."

I set and released the brake a couple of times, undecided. Then I got off the bicycle and walked over to the almonds. I grabbed them, a handful at a time, and filled the basket to the brim. Before taking off, I hid the Walkman among the shells, with a piece of colored tape on the play button.

BY THE TIME my mother found the box with the almonds it was already February, maybe March. She had taken advantage of my being at school to come in and straighten up my room. She was constantly moving things, throwing stuff away. She left the box on the bed. When I came home and saw it, I felt, strangely, that I had neglected something important. I opened the box; it was empty. I ran my index finger over the bottom, where a fine powder had been deposited, and swallowed it with my saliva. It wasn't sweet, it had no taste, yet it made me picture Bern again, bending over the shells, and for the rest of the day I couldn't concentrate on anything else.

But that afternoon was an exception. In those early years, toward spring, Speziale and the masseria began to seem unreal. I forgot about Bern and the others until it was time to go back there, in August. I didn't know if it was the same for them. If they missed me, they certainly didn't show it. When we saw one another again, we didn't brush cheeks or touch hands, we didn't ask about the months that had passed. They treated me as if I were merely another element of nature, a phenomenon that appeared and disappeared according to the seasons.

As I got to know them better, I learned that for them, time had a different flow than it did for me—or rather it didn't flow at all. Each day was marked by three hours of theoretical instruction in the morning, followed by three hours of manual labor in the afternoon, with the sole exception of Sunday. That rhythm remained unchanged, even in summer. So I stayed away from the masseria before lunch, preferring not to get drawn into one of Cesare's lessons. They made me feel stupid. He spoke about the myths of creation, about wedge or splice grafting on fruit trees, about the *Mahabharata*, all things that I knew nothing about.

Every so often the boys went off with him alone, one at a time. They would sit and talk together in the shade of a large holly oak. Actually, it was always Cesare who spoke, while Bern or Tommaso or Nicola nodded their heads up and down. One day, Cesare told me that I was welcome to talk a bit if I wanted to. I thanked him, but I never had the courage to follow him under the tree.

And yet, year after year, I, too, became part of the masseria. The summer following my first year of high school, and the one after the second. My father wasn't thrilled about it, but he didn't say anything, because knowing that I was at the neighbors' place was better than seeing me moping around the house looking sullen all day.

In return for the hospitality, I contributed to the chores as best I could. I picked green beans and tomatoes, I pulled up clumps of weeds from the dirt track, and I learned to weave dry branches into garlands. I was all thumbs, but no one criticized me. When my braid got so tangled up that I couldn't go any further, Bern and Nicola rushed to help me. They undid the work to the spot where I'd gone wrong, then explained the sequence yet again: take that end, slip it under here, then in between, now tighten it, there, you can continue. They could have knotted those branches with their eyes closed, made miles-long garlands, even if they served no purpose: as soon as they were done, they burned them. When I asked Bern why they wasted so much time making them, then, he replied: "It's for humility. Just an exercise."

I remember one evening when we were all under the pergola, where bunches of black grapes hung over our heads. Nicola was lighting a fire in the brazier while the other boys carried the dirty dishes into the kitchen. I had barely tasted the food. They were all vegetarians at the masseria, and at the time I hardly ate any vegetables at all. But I endured my hunger just to stay there, in that peaceful stillness, removed from everything, close to Bern and the fire.

Cesare entertained us with the story of when, at the age of twenty, he'd had a vision of his previous life.

"I was a seagull," he said, "or an albatross, anyway, a creature of the air. I flew across the sky, all the way to the shore of Lake Baikal."

He challenged us to find the lake on the tablecloth's map. The boys scrambled to push aside whatever still cluttered the table and began furiously scanning the continents.

It was Nicola who shouted first: "Here it is! Right here!"

Cesare rewarded him with a taste of liqueur. Nicola sipped it triumphantly, as Bern and Tommaso glowered. Especially Bern. He stared at the tablecloth, at the blue speck of Lake Baikal, as if he wanted to memorize every single name on the map once and for all.

Then Floriana served ice cream and calm was restored. Cesare went on talking about previous lives, those of the boys this time. I've forgotten what he said about Nicola; he said Tommaso had been a feline and that Bern had retained something subterranean in his blood. My turn also came.

"And you, Teresa, my dear?"

"Me?"

"Which animal do you feel you have been?"

"I don't know."

"Try to imagine, go on."

They all looked at me.

"I can't think of anything."

"Close your eyes, then. And tell me the first thing you see."

"But I don't see anything."

They looked disappointed. "I'm sorry," I mumbled.

Cesare stared at me from across the table. "I think I know," he said. "Teresa has been underwater for a long time. She learned to breathe without oxygen. Isn't that so?"

"A fish!" Nicola exclaimed.

Cesare looked at me as if he were seeing right through my body and through my past lives.

"Not a fish, no. An amphibian, if anything. Let's see if I'm right."

The boys could tell that another contest was coming, and they lit up instantly.

"On my count of three, you must all hold your breath. The one who lasts the longest wins."

He counted slowly, at two I filled my cheeks with air and froze. We eyed one another, and no one giggled, as Cesare crept behind our chairs and stuck a finger under our nostrils to make sure we weren't cheating.

The first to give up was Nicola, who got angry and stood up, disappearing inside the house. After him, Bern. Then Cesare stationed himself between Tommaso and me, checking us in turn. My throat started to convulse, but Tommaso, whose neck was a disturbing violet color, opened his mouth an instant before me.

Cesare offered me the glass of liqueur that I had earned. I drank it too quickly and the alcohol's warmth exploded in my stomach. It was all so serious, so solemn, as the others watched me drink, as if that act finally marked my coronation as an honorary member of the family: the masseria's first sister. I didn't admit that for days I'd been practicing holding my breath in the pool, one of the games I played by myself. It was far more fascinating to believe in a past life, when I was like the frogs that had sprung up in the countryside two summers ago. I could choose what to believe in. I didn't know that before I went there.

That year Bern lent me a book. He said it had intrigued him, that it seemed to be talking about him. As I turned the volume over in my hands, I felt that he was looking at me differently, as if he were observing a rough stone and wondering if it was really worth polishing, whether it would survive the transformation or would prove too fragile.

At home I put the copy of *The Baron in the Trees* on my bedside table. My grandmother noticed it. "Did they assign you Calvino to read over vacation?"

"No."

"So you chose it yourself?"

"More or less."

"You'll find it difficult."

The next few hours I took the book with me everywhere, out to the yard, to the pool, but I never opened it. That night, in bed, I tried but I lost my concentration immediately.

A few days after lending it to me, Bern asked me if I'd liked it.

"I haven't finished it yet," I said.

"But did you get to Gian dei Brughi? It's my favorite part."

"I don't think so. Maybe I'm almost there."

We were walking along the dirt track. It was a sultry evening; disco music reached us from afar.

"And the swing?"

"I don't think so."

"Then you haven't read a word!" he said, scowling. "Give it back to me right now!"

He was shaking. I begged him to let me keep the book for a few more days, but he demanded that I go and get it. Once I did, he walked away without saying goodbye, pressing the book to him tightly.

As he disappeared into the darkness, I felt a twinge of sadness, as I often did toward the end of my time there. My thoughts became repetitive: this is the last time you'll wear your bathing suit, this is the last time you'll watch the cat slink over to the pool, this is the last time you'll leave the masseria behind. This is the last time you'll see him.

THEN, THE FOLLOWING SUMMER, I was seventeen; Bern had turned eighteen in March. A bed of reeds had grown up at a spot in the countryside where water gushed from an underground spring. It was a ten-minute walk from the masseria through the olive trees. Bern took me there during the hottest time of the day, while the others slept: our secret hours from the beginning.

We lay down on the ground and I closed my eyes. Suddenly the color I saw printed on my eyelids changed; I thought it was because of a cloud, but when I opened my eyes I saw Bern's face very close to mine. He was breathing hard and looking at me gravely. I nodded almost imperceptibly and he bent down to kiss me.

That day I let him stroke my face and run his fingers over my hips as we kissed, nothing more. But we always wore so little at Speziale, and the reed bed was so remote from everything. We went back every afternoon, always venturing something more daring.

The soil along the brook was muddy, I felt it stick to my back, my hair, the soles of my feet, and Bern's body over mine also seemed to be made of clay. I clung to his back with one hand and sank the other hand into the earth, among the stones and the worms. From time to time I looked up: the reed stalks seemed very tall.

During that August, Bern explored every inch of my body, first with his fingers and then with his tongue. At times I was so confused, so consumed with excitement that I no longer knew where his head, his mouth, his hands were. I gripped his hot erection and at first I had to help him shove it between my legs, because he seemed paralyzed by fear. I had never been with a boy, and in a single summer he took everything there was to take.

Afterward he wiped away my perspiration with his hands. He blew on my forehead to cool me. Then he wet his thumb with saliva and rubbed the muddy smears off my skin, brushed the leaves out of my hair, one by one. We always had to pee and we did it side by side, me squatting and him kneeling. I watched the rivulets of urine make their way along the ground and hoped that they would join together; sometimes they did. Then we would go back to the masseria, not holding hands, not speaking.

At first I was afraid that he would tell Cesare everything during their talks in the shade of the holly oak, but their relationship had changed over the past year. Throughout the whole summer I did not witness a single prayer, except for the brief blessings before meals. There was no singing, and no lessons. Beginning in September, Bern and Tommaso would start attending a school in Brindisi, to prepare for the high school diploma exam as Nicola had done the previous year.

By then the four of us spent a lot of time outside the masseria. We waited for the cooler hours because of Tommaso's pale complexion, then we'd pile into Floriana's Ford. There was a narrow cove at Costa Merlata, where we would lie on the flat concrete area that served as a beach. Nicola and Bern dived from the highest point of the rocky bluff. From below, Tommaso and I would assign them marks. We didn't know what to say to each other. Some tiny fish kept nibbling at my heels and ankles; I kicked my feet to chase them away, but a second later they were back.

Then Bern and Nicola joined us, swimming. Bern held me up with one hand and furtively shifted the edge of my bathing suit aside with his fingers, all the while talking to the others.

In the evening we went to the Scalo. A cooperative of young people had taken over a stretch of rocks by the sea, near an abandoned watchtower. A few benches and tables were scattered around a pink trailer. Bern and the other boys knew everyone there and were constantly greeting people. I almost always ended up off to the side, sipping a beer, alone or with some wasted-looking stranger.

One night I was stunned to see Bern and Tommaso devour a sandwich with chunks of horse meat. I was sure that for Cesare eating horse meat was a very serious infraction. Nicola picked at his fries unconcernedly, as if by then he was used to their behavior. But when Bern wiped the ketchup off his lips with the back of his hand and told him that someday he was going to tear apart one of his father's nice plump hens, Nicola leaped to his feet, challenging him with his height. Bern and Tommaso teased him, flapping their elbows up and down like a couple of chickens.

Toward midnight we walked back to the car on the path through the myrtle bushes, each clinging to the shoulders of the person in front of him.

When we got back to the villa, the boys got out of the car to walk me to the door. The pool was inviting at that hour; we joked about diving in, clothes and all, and how my father would throw stones at us, but we never did it. From my bedroom window I heard the Ford's engine start up. My hair was frizzy with salt, my fingers stank of cigarettes, my head was dizzy from the beer, and I had never felt so alive.

AFTER A WHILE the reeds were no longer enough for us. Bern became obsessed with the idea of a real bed. If I asked him what was so different about it, he answered me vaguely: "You can try a lot more things."

But we didn't see how we could manage it: at the masseria Cesare was always around, and at the villa Cosimo and his wife Rosa stood constant guard. We considered all the possibilities over and over again.

Meanwhile, we were already past the feast of San Lorenzo, and the heat was different, summer was easing up a little. Everything around us transmitted a sense of urgency.

"I'll come at night," Bern said finally, as his fingertips traced circles around my navel.

"Where?"

"Your place."

"They'll hear you. Nicola always says he sleeps more lightly than everyone else."

"It's not true, I'm the lightest sleeper. And besides, Nicola isn't a problem."

"What if my father hears us?"

Bern turned his head. His eyes were so close to mine, almost unbearable.

"I don't make any noise," he said. "You're the one who should restrain yourself."

But it was another few days before we acted on the plan, days when we did not return to the reed bed because Bern was too focused on the details. I was sorry about it, but I didn't tell him, just as I never told him that I had fallen in love with him. Though I did my best to drive away the suspicion that getting to the bed had become more important than being with me, that doubt haunted me more and more each afternoon, when he took my hand and instead of leading me beyond the oleanders, he took the dirt track.

From a hidden spot, we studied my grandmother's house. "I can put one foot on that ledge, then grab on to the eaves," Bern said. "Did you try to see if they'll hold? From there I should be able to reach the windowsill, but you'll have to help me. Come to the window when you hear this sound," he said, sucking in his lower lip and whistling something that sounded like a bird's call.

On the appointed evening, we did not go to the Scalo. Bern told the others he didn't feel like it, after all, we'd spent every night there, couldn't we come up with something different?

"Such as?" Nicola asked him, a little annoyed.

"Such as buying something to drink and bringing it with us to the piazza."

He always got his way, Bern did, so we went to Ostuni. In Piazza Sant'Oronzo children were running around everywhere; we sat down in the center of the square, at the foot of the saint's statue. It was ten days to the patron saint's feast day, but the lights were already up.

"My father asked me if there would be other girls with us," I said.

"And what did you say?" Tommaso asked.

"'Of course,' I told him."

My back was leaning against Nicola's knees, my legs were stretched out over Tommaso's, and Bern's head was against my shoulder. We had bought a large bottle of beer, because it was more economical, but mostly because we liked passing it from hand to hand, drinking and exchanging our saliva. I felt more connected to the boys than ever, and I liked it. And besides, there was the secret, what Bern and I would do that night.

When we went back to the parking lot around one a.m., a group of guys were standing beside the Ford; they had set their bottles on the roof. Nicola told them to remove them, a little brusquely maybe, but not enough to justify the tone with which one of them asked him to repeat it, this time saying "please."

Bern blocked my way. I saw Nicola pick up the bottles and one by one move them to the guys' car. The group hooted in concert, making fun of his bravado. Bern didn't move, his right arm still stretched out protectively to keep me from moving forward.

Then one of the guys in a red surfer suit and spotless Nikes offered Nicola a beer.

"Chill, dude," he told him. "Have a little drink." Nicola shook his head no, but the other insisted. "As a peace sign."

Nicola took a sip and handed it back to him. He opened the Ford's door. It would have ended there, he would have backed out and we would have gone to join the snaking line of cars headed to Speziale, if another guy hadn't pointed to Tommaso, saying: "Did they dip him in bleach?"

Nicola, lightning swift, dealt him an openhanded blow to the face. It was the first time I'd seen anyone strike a person like that. I gripped Bern's arm; he hadn't taken a step, as if he had foreseen it all the very moment we'd arrived there.

Nicola leaned over the guy and whispered something that none of us heard. Then they were off.

We got in the car, Tommaso and I in back, Bern and Nicola up front. When we were on the road again, stuck in the line of traffic, all three started shouting excitedly.

At home, I found my grandmother in the living room. She had fallen asleep with the TV on. I touched her lightly and she started.

"Where have you been?" she asked, rubbing her cheeks.

"Ostuni. At the piazza."

"Ostuni is terrible in the summer. All those loud tourists. Would you like some herbal tea?"

"No, thank you."

"Then make some for me, if you would."

When I brought her the cup, she was still motionless as I had left her, her eyes wide open, staring at the screen.

"Is it the dark-haired one?" she asked, not turning her head.

The cup clinked on the saucer. "What?"

"Yes, it's the dark-haired one. The other one, the real son, is also cute. But the dark one is without a doubt more captivating. What's his name?"

"Bern."

"Just Bern? Or Bern as in Bernardo?"

"I don't know."

She was silent for a moment. Then she said: "I was trying to remember what we used to do at night when I was your age. And do you know what we did? We went to the piazza, in Ostuni. Is he kind to you?"

"Yes."

"Good."

"I'll bring your tea up to your room," I suggested. "That way you can lie down." She followed me up the stairs. Before leaving her to herself, I added: "Don't tell him, please."

I took her smile to mean she wouldn't. In the hallway, I stopped in front of my father's door and heard his heavy breathing.

I took a shower, then more time went by, a period during which I took off my night socks and put them back on, tried on at least four different sleep shirts, lay down under the sheet, and then moved to the chair because Bern might not like getting into a warm bed. The thought of what, at the reed bed, had come naturally now made me nervous.

By three a.m. I convinced myself that he wouldn't come. Maybe he hadn't been able to get away, or else he had forgotten. I focused on the second hypothesis. Yes, the brawl we had come close to had caused him to forget our plan.

But after a while, I heard something thump. I imagined his foot on the eaves. I forced myself to stay where I was until he whistled. When he reached the window, I opened the shutters and helped him climb in. Right away, he kissed me. His breath smelled of beer. His hands groped for my breasts, first through the shirt, then pushing it up, out of the way.

"You're tense," he said, continuing to touch me.

"I'm afraid they'll hear us."

"They won't hear us."

He pulled away to glance at the bed against the wall.

"Do you want to stay on top of the sheet or under?"

"I don't know."

"I'd rather be on top. What about the lamp? Should we leave it on?"

We knelt on the bed, facing each other. He had undressed too. It took my breath away to see him like that, naked in the dead of night, his erection rising from the dark patch of pubic hair.

He pushed toward me with the same frenzy as before, but this time I stopped him. I told him that we would do it differently, that we would do it slowly. We were in bed and we had all the time we wanted. He pulled back, looking confused. So it was I who went to him and made him lie back as I put my knees around his waist.

I started rubbing myself back and forth, from his stomach to his legs, back and forth, slowly at first, then faster, until I felt something forming at the point where we were in contact, a kind of heat that quickly rose to my throat. It had never happened to me before.

Bern stared at me in astonishment, his hands grounded on the sheet, as if he were afraid of interrupting what I was doing.

My first thought, afterward, was that we had been too loud, that maybe I'd cried out or he had.

"It was different than I thought," he said. "You didn't even let me move."

"I'm sorry."

"No," he said hurriedly. "It was good."

My forehead was resting against his collarbone, I felt like going to sleep, but I could feel that his muscles were still tense.

"I have to go now," he said.

From the bed I watched him put his clothes back on. I wasn't embarrassed to lie there naked, what embarrassed me was to still want him while he was getting ready to go back to the masseria.

"You can leave by the door," I said.

But he was already climbing through the window. I went over to it. He had gone down a foot or so when he looked up one last time.

"Did you see how great Nicola was? He protected us all."

He steadied a foot between the stones of the façade and jumped down. When he reached the pool, he waved goodbye to me, then started running.

THE FOLLOWING DAY my father asked me to go to Fasano with him to see a childhood friend of his. I didn't want to go, but I felt guilty about what had happened the night before, so I said yes.

The friend lived in a yellow row house in the suburbs. He was very fat, and had trouble breathing; he didn't move from his chair the whole time. With him was a girl my age; she brought him water if he was thirsty, picked up the pillow that was always falling to the floor, and at one point she lowered the shutters a few inches because she noticed that the light was bothering him. She performed those tasks with detachment, as though far away, then sat back quietly and listened to the conversation, or more likely did not listen at all. I found myself staring at her thin, tanned legs.

My father's friend coughed constantly into a crumpled handkerchief, then examined it for traces of something. I asked permission to go outside and get some air.

After a few minutes, the girl joined me. I was smoking a cigarette behind the wall.

"I have some grass, if you want," she said.

She pulled a small plastic envelope out of her pocket. She asked me for a cigarette, then with precise gestures emptied out the tobacco. Her fingernails were painted, the polish several days old. "Do you know how to make a crutch?" she asked. While I made it, she mixed the grass with the tobacco. She lit up the joint and we took a few tokes each.

"Is it serious?" I asked her.

The girl shrugged as she blew on the tip, making it gleam red. "He'll die, I think."

I told her my name and somewhat awkwardly held out my hand.

"I'm Violalibera," she replied.

"What a beautiful name."

She grinned self-consciously, forming two dimples. "I had another name, but I didn't want it anymore."

"What was it?"

She looked aside for a while, undecided. "It was Albanian," she answered finally, as if that were enough.

I didn't know what else to say, I was afraid I'd been intrusive, so I asked her: "Do you ever come to the Scalo?"

"What's that?"

"Kind of an outdoor club, by the sea. They show films. And there's a bar, but it only sells beer and horse-meat sandwiches."

"How disgusting."

"They're a little greasy. But you get used to it."

We finished smoking, both of us absorbed in our thoughts.

"Can you give me some?" I asked her. Bern and the others would be thrilled. They often mentioned the idea of buying some grass, but they never had the money. "I'll pay you."

Violalibera pulled out the little envelope. "Keep it. I have more."

She stuck a piece of candy in her mouth and offered me one. Then we went back into the house and she served us almond milk. The friend had a coughing fit that caused the milk to go down the wrong way. My father went over to him, but didn't know how to help him. Violalibera told him not to worry, and slapped the man on his back until he stopped coughing; then she brought the tray with the pitcher back to the kitchen. For the rest of the time I had to keep my chin pressed against my chest so I wouldn't burst out laughing.

My father was sad on the way back. He asked me if I felt like taking a little walk on the seafront promenade, maybe have an ice cream. I was anxious to get back to the masseria; there were only a few days left and I was wasting all that time, but I didn't want to let him down.

We ended up on the beach of Santa Sabina. The sand was firmly packed, fishermen's boats swayed near shore. My father took my arm.

"Giovanni and I used to come fishing here as boys," he said, pointing to an indefinite spot offshore. "We would return home with buckets full of fish. You could do that then, there weren't all the rules we have now. Whatever you caught was yours."

He rotated the cone in his fingers, smoothing the ice cream with his tongue.

"I'd like to come back to live here one day. What do you think?"

"I think Mom would never agree to it."

He shrugged. At the end of the pier was a darkened amusement ride, the seats tied together by a chain.

"Giovanni knew your friend's father."

"Cesare?"

"No. The father of the other boy. Bern, right?"

He was looking at me very closely. Had my grandmother told him about Bern? I hoped he wouldn't say anything else, but he went on: "They called him 'the German.' Nobody knows where he ended up."

"Bern's father is dead. He told me himself."

He winked at me. "He doesn't give me the impression of being very sincere."

"Should we head home, Dad?"

"Wait a minute. Don't you want to know why he's called 'the German'? It's a curious story. Have you ever heard of the tombaroli?"

I recalled seeing a short paragraph on a page in a history book. I kept silent.

"It's full of underground tombs around here. You can find anything in there: arrowheads, obsidian, ceramic shards. As a boy I picked up a thing or two myself. As I told you, at that time whatever you found was yours. But for the German and his friends it was different. They came here on vacation and instead of going to the beach they devoted themselves to archaeology. So to speak."

He wiped his sticky lips and fingers with the ice-cream napkin, then crumpled it and tossed it away.

"They dug at night. When they'd filled up, the German loaded everything into a van and went to sell the goods in Germany. He'd made good money. One year he arrived in Speziale in a Mercedes. The carabinieri went looking for him. So you know what he did? He emptied a huge tomb all at one time, then he left and never came back. It caused quite a stir in Speziale, you can just imagine. Giovanni says everyone talked about it."

The seagulls didn't budge as we walked by. They screeched and flapped their wings nervously.

"Let's go home, please," I said in a whisper.

ALTHOUGH I DIDN'T WANT to admit it, that story left me troubled. When I was in the reed bed with Bern again, I couldn't let myself go. The roots scraped my back and the mud on my elbows bothered me. I felt as if there were a thousand eyes on us.

A fighter-bomber tore through the sky over the tops of the reeds. Then there was a rustling sound, and when I snapped my head up to look, I saw the grass swaying and heard footsteps hurriedly moving off. I told Bern, but he didn't pay any attention to me.

"It must have been a cat. Or you imagined it."

Later the others found us under the pergola, pretending to wait for them to play skat as usual. Tommaso barely said hello to me. By then he and I spent all our time competing for Bern's attention.

After a few minutes Cesare also appeared. He smiled at me distractedly, then turned to the boys: "The chicken coop has to be cleaned out. Who's coming to help me?"

Bern and Tommaso exchanged dark looks, pretending not to have heard him. With a kind of resignation, Nicola said, "I'll be there in a minute."

Cesare stood there waiting a few more seconds. Then he nodded to himself and left.

Bern set down a winning combination of cards. As he reshuffled the deck, I thought about the way he pronounced all the other German words in the game. He must have learned them from his father, I imagined. Then I forced myself to chase away that suspicion.

THAT YEAR my departure coincided with Tommaso's eighteenth birthday. On the last night we had a lot to celebrate, a lot of reasons to drink ourselves into a stupor.

We'd brought a bag with a change of clothes to the beach and I undressed behind a low wall. I put on a pair of rope sandals, a skirt I'd bought in the spring with my mother, and a top. The fabric was slightly itchy on my salt-caked skin.

I remember what the others were wearing as well: Tommaso's mustard-yellow tee, Bern's black one with ZOO SAFARI on it, which he would still have ten years later, and Nicola's flashy shirt. I remember my anguish, which grew stronger hour by hour, at the thought of leaving the following morning.

When we reached the Scalo, the sky was pink. I showed the boys Violalibera's grass; although Nicola wanted to try it right away, we decided to save it for later. Nicola and Bern had planned a surprise for Tommaso: they had set aside a bottle of gin and some pineapple juice. We mixed them in a jug. The cocktail was so strong that in less than half an hour we were plastered, sitting on the deck chairs in the dark.

A screen mounted in the middle of the patio area projected images of a black-and-white film whose actors seemed to move jerkily. I had realized immediately that Tommaso's birthday would make everyone forget that I was leaving, and I decided that before the night was over I would have to make Bern kiss me in front of the others. What would I return to Turin with otherwise?

We went off to smoke and each of us expressed a wish for Tommaso now that he had turned eighteen. I wished for him to find a girl soon; he thanked me, but with a sort of sardonic grin. Bern spoke last and said, "May you learn to dive from any height."

But toward me Bern remained remote, distracted. He and Nicola drank toasts only to Tommaso, then they lifted him into the air by his armpits. The pineapple juice was finished, so we had stopped diluting the gin. The bottle fell into Tommaso's hands and stayed there. He took long swigs that left him gasping.

Then Bern decided we had to climb the tower, he wanted to show me something. Nicola stayed behind, he'd already been there, he said, while Tommaso joined in reluctantly. He just didn't want to leave us alone, I thought.

We approached the barbed-wire fence that surrounded the ruin. The light in the distance was barely enough to read the DO NOT ENTER sign. Bern removed a post to create an opening. We had to walk through a stretch of nettles and my legs were bare; I told him I would get pricked all over, but he kept going straight ahead.

We scrambled up the stairs to the center of the tower. A small window opened onto the sea, but all it framed was a dark rectangle. Bern lit a flashlight. "This way," he said.

We took a flight of steps, downward this time. The walls were covered with graffiti and carvings, and the ground was littered with glass shards that crunched under our sandals. Beads of sweat began to run down my body. I asked Bern to turn back, but he said he wanted to take me to the bottom.

"I don't want to, let's go," I whimpered.

"We're almost there."

Behind me, I smelled Tommaso's alcoholic breath. I clutched Bern's T-shirt, tugging at it, but he kept going down.

Then the steps ended. We were in a room; I didn't know how big it was until Bern aimed the flashlight around.

"Here we are."

The light beam picked out a mattress tossed on the floor in a corner. Around it were empty bottles and cans arranged in an orderly fashion. He bent down to pick one up and showed me the date on the faded label.

"Look at the year: 1971. Can you believe it?"

Even in the darkness his eyes glittered excitedly. But I wasn't the least bit interested in the can, or any of the rest. I imagined cockroaches crawling in the dark, near my feet.

"Let's go," I begged.

He put the can down.

"Sometimes you act like a spoiled brat."

Though I couldn't see Tommaso, I had the impression that he was smirking behind my back.

Bern quickly took the stairs, leaving me behind. I stretched my hands out in front of me to avoid bumping into walls that suddenly appeared before me. When we got outside, I threw up my supper on the nettles. Bern said nothing and did not move to help me. With his thumb, he kept switching the flashlight off and on. He looked at me coldly, as if he were assessing me. Only when it was time to duck under the barbed wire did he offer me his hand as support, but I didn't take it.

Meanwhile, the Scalo had filled with people. We started dancing. I felt increasingly left out of the evening's excitement, but I fought against it, determined not to let sadness ruin the last moments. The music of Robert Miles was playing, melancholic and dreamy; I wished someone would change it or else that it would go on forever. I was of two minds about everything.

It was while we were dancing that Tommaso threw himself at Bern, his forehead against his stomach, and started sobbing. Bern took Tommaso's head in his hands, bent down to whisper something in his ear. Tommaso shook his head firmly without breaking away.

"Come with me," Nicola told me.

We ordered a couple of beers. I thought about the effect that mixing the grass with all that alcohol would have, about how I would face the car trip the next day, and then I thought, What the hell. Bern and Tommaso were still in the middle of the dance floor, but Tommaso had gotten to his feet and now they were in each other's arms, as if they were doing a slow dance.

"What's come over him?" I asked Nicola. He lowered his eyes and said: "He's just had too much to drink."

In another month Nicola would start attending university in Bari. Throughout the summer it had been as if that prospect, that year of advantage over the other two, kept him somewhat at a distance.

"It's after three," he said, "we have to go home, Cesare will be beside himself. And your father, too."

Tommaso and Bern had wandered off toward the sea, I saw them sit on the reef and then lean back on the rock, as if waiting for the tide to catch them.

"We'll wait for them," I said.

Nicola tried to lead me away by the arm, but I pulled free and ran to Bern. His head and Tommaso's were close, but they weren't talking, they were simply gazing at the dark sky.

When Bern saw me, he stood up with a kind of submission, as if he knew he would have to. We took a few steps, even farther into darkness.

"I'm leaving," I said. I was unable to contain my anguish, my whole body was shaking.

"Have a good trip tomorrow."

"Is that all you have to say to me? Have a good trip tomorrow?"

Bern glanced at Tommaso, who had not moved. Then he took a deep breath. Suddenly I was certain that he was in full command of himself: the grass and the gin had not affected his lucidity, not even for a moment.

"Go back to Turin, Teresa. Go back to your home, your classmates, your comforts. Don't worry about what goes on here. When you come back next year, nothing will have changed."

"Why don't you ever kiss me in front of the others?"

Bern nodded, twice. He had his hands in his pockets. Then he moved toward me and grabbed my hips.

It was not a hurried kiss, nor an awkward one. On the contrary, he pulled my body to him so that it fit tightly against his own. He slowly ran one hand up my back and grabbed my hair. But it was like kissing someone else, someone I didn't know at all. It was, I thought at that moment, the perfect simulation of a kiss.

"I suppose that's what you had in mind," he said.

Tommaso kept his eyes closed, but even so, he was present between us. Bern stared at me, not angry, but rather with regret, as if I were already in a car speeding off, unreachable behind the window. I backed away, still looking at him, before turning around and running. I left him with the tower's ruins behind him, the foam-slick rocks, the silent sea, and, all around, the mercilessly bright night of the South.

BY THEN I was used to finding Turin more inhospitable than I had left it, the avenues too wide, the sky white and oppressive. Cesare had once said: "In the end, everything that man has constructed will be reduced to a layer of dust less than an inch thick. That's how insignificant we are. It is only the thought of God that makes us worthy." Amid the buildings in the city center his words came back to me; everything seemed precarious, meaningless. Even so, I sensed that within a week or two the vortex that had formed in my chest, something midway between hunger and nausea, would dissolve and everything would return to normal. That's how it had always been. But that year the sadness lasted much longer. At Christmas, I was still torn by longing for Speziale.

In June, I turned eighteen. On my birthday I found an envelope on the pillow with some cash and a card with a heart, drawn in pen, with the number eighteen in the middle. I put the money between the pages of my French dictionary, then waited all day for a phone call from Bern, which did not come. Yet I had told him the date, I had even written it in a letter sent a few weeks ago, to which he had not replied.

My grandmother called instead. She was unprepared when I asked her about Bern, Tommaso, and Nicola. She repeated the same words she had used once before, "They come and go." I thought she did it deliberately.

Final grades were posted on the bulletin boards with no surprises, but I didn't feel like celebrating. My friends left on a trip to Spain, and I was able finally to devote my time to counting the days that separated me from Speziale.

In a single afternoon I spent all the money I'd stashed in the dictionary. I bought a bikini and gave the rest to a Tunisian guy in exchange for some weed. When I got home, I hid the weed between two carved-out halves of a soap bar, as he had advised me. Bern had sworn that everything would be the same as the year before.

THE LAST STRETCH of highway, after Bari, ran past a series of nurseries. Palms set out in rows swayed beyond the fences. They had always been the sign that Speziale was getting close. That year I found them decapitated, the trunks lined up like the teeth of a rake. I asked my father what had happened, and he glanced over distractedly.

"I don't know," he said, "they must have pruned them."

The two palms at the entrance to the villa were also dead. Cosimo explained that it had taken an excavator to yank out the roots.

"Come, I'll show you one of the little bastards," he said.

He invited us to follow him into the lodge, but only I went. From a shelf that held piles of tools, he grabbed a glass jar. On the bottom lay a red beetle, with a long curved proboscis.

"The red weevil," he said, shaking the jar under my eyes. "It burrows into the bark and lays its eggs. Thousands of larvae are born from just one egg. They eat the palm from inside, and when they're finished, they move on to another one. China sent these damned pests to us."

During the next few hours, I tried not to rush over immediately to see Bern. In the evening I lingered with my grandmother and my father on the terrace, going on at length, telling her about the school year, until I was bored by my own voice. My back was to the railing, but as soon as I got up to help with the dishes I glanced in the direction of the masseria; beyond the foliage of the olive trees I glimpsed a very faint, yellow gleam of light, which seemed to be an infinite distance away.

In the morning the sky was milky. I told my grandmother that I was going for a walk and that I might stop by to say hello to the boys. I was wearing the new swimsuit under a white sundress and I hoped she couldn't see how I was shaking. In the straw bag was the soap bar with the weed; I would give it to Bern right away, partly to surprise him and partly because keeping it at home was too risky, with Rosa touching everything.

But my grandmother stopped me. "Breakfast first." A sour-cherry croissant waited on the table next to a glass of milk. I hesitated, then I perched on the edge of the chair and she sat down across from me. I broke off a piece of croissant with my fingers and chewed it.

"Good?" my grandmother asked.

"You know it's my favorite."

Now I'd have to go back into the house to brush my teeth again and I'd lose more time.

"Well, enjoy it. You won't find these in Turin."

One of her books was lying on the table. I turned it over to look at the cover. *The Skull Beneath the Skin: A Cordelia Gray Mystery*.

"Are you enjoying it?" I asked, just to say something.

She waved her hand. "I just started it. It's not bad."

"Do you always guess who the killer is?"

"Almost always. But sometimes these novels trick you, you know."

There must have been a cicada hidden somewhere close by, because every time I moved, it suddenly fell silent before resuming its tireless chirping.

Not far off, Cosimo was tinkering with the irrigation system. He stood in the center of the spray with his arms folded.

I finished the croissant in silence and drank the milk. My grandmother started fiddling with a corner of the book jacket.

"You won't find him at the masseria," she said finally.

"What?"

I had some greasy crumbs stuck to my fingers, but there was no napkin. I wiped them on my legs.

"Bern. You won't find him."

I rested an elbow on the table. Although the sky was overcast, the light was strong, it hurt my eyes. The buttery taste of the croissant rose from my stomach as a burp, which I held in. My grandmother let go of the book and reached her hand toward me, but I drew back.

"Remember when you asked me about him, on your birthday?"

"Yes."

"It was true that I hadn't seen anyone from the masseria for some time. Bern and that other boy . . ."

"Tommaso?"

"No, not Tommaso. Yoan."

"There is no Yoan."

"Maybe you didn't get to meet him. He arrived at the end of last

43

summer. They worked here in December, he and Bern, for the olive harvest. The oil turned out to be exquisite. But you tried it, of course, I sent it up to your—"

"And then?"

My grandmother sighed.

"After the harvest there wasn't much to do, so I didn't call them. But a few weeks ago I was curious to know how they were. Bern had told me he was having some problems in mathematics. I had offered to help him and I felt guilty about not having asked him about it again. So I went to the masseria. It was already July, I think. There was only Cesare's wife, Floriana, and from her I learned . . . well, about what happened."

I saw my father pop out from behind the house. Seeing us there, he quickly disappeared again.

"What happened, Nonna?"

"It seems that Bern got into some trouble," she said, looking steadily at me, "with a girl."

With my finger I picked up the remaining crumbs one by one. Instinctively, I put my finger in my mouth and sucked it.

"What kind of trouble?"

My grandmother smiled sadly. "The only kind of trouble you can get into with a girl, Teresa. He got her pregnant."

I jumped up abruptly. The chair fell over backward, clattering onto the stone. My grandmother flinched. "I'm going to see," I said.

It didn't even occur to me to straighten the chair back up.

"You can't go there."

"Where's the bike? Where the fuck did you put it?"

I FOUND the iron bar across the dirt track padlocked. I dropped the bike on the ground and ducked under the bar. On my right I noticed a tree laden with yellow pears, many of which had fallen.

There was no one at the masseria. I sat on the shaky swing-chair, without swinging.

So Bern got a girl pregnant.

I watched the cats slinking along the walls. A gigantic one with a reddish coat stared at me for a long time.

Bern got a girl pregnant. Why wasn't it me?

I didn't move when I heard the sound of an approaching car. Cesare and Floriana were dressed for town, he in a blue cotton suit and tie, she in a floral-print sheath dress. Behind them walked a boy with his head down, he, too, dressed up, but without a tie. Cesare had cut his hair. I wanted to run to meet him, but instead I remained still.

"Teresa, my dear," Floriana said, taking my arms and swinging them open, as if she wanted to see all of me. "We were at mass. Were you waiting long? With this heat. I'll run and get you a glass of iced tea right away."

"I'm fine, thanks."

My heart was out of control. I was afraid she could feel it pulsing in my wrists.

"But of course, a little iced tea to cool off. I made it yesterday. And I used agave instead of sugar, so you don't have to worry about your figure. You haven't met our Yoan, have you?"

With that she quickly disappeared inside the house. Yoan gave me a kind of bow, not saying a word, then he too went off. Cesare loosened his tie, puffing because of the heat. He took a chair from the table and placed it in front of me.

"We found this parish," he said. "It's a little far away, in Locorotondo, but the priest is open to ideas, and I believe he thinks highly of me. Don Valerio. He's doing a nice job with Yoan. The boy is Orthodox, even if he doesn't really know what it means. I'd like for you to meet him, Don Valerio, that is. Are you just passing through, or are you staying awhile this year too?"

Something in the way he spoke made me suffer more intensely.

"You're not lucky with the weather," Cesare was saying, "perfect until yesterday, but now . . . too humid."

"I came to say hello to Bern."

Not to seem rude, I added: "And Nicola."

Cesare slapped his palms on his knees. "Oh, Nicola, that blessed son of mine! Since he's been at university we see very little of him. But he's doing well, I must say. He took all the exams except one."

"And Bern?"

Cesare didn't seem to hear me. He was trying to remove a stain from his shirt with a saliva-moistened finger. His beard had disappeared too, that's what else was different. His round face, clean-shaven like that, had something childlike about it.

"Nicola arrives in four days," he said, "he'll be here a week. I think he will have to study, he always says he has to study, but I'm sure he'll be pleased to see you."

Floriana returned to the yard with a glass of iced tea. The rim was white with lime; it wouldn't have bothered me under other circumstances, but at that moment I decided that I would not put my lips to it. Every detail struck me as a new betrayal: Cesare's appearance; Floriana, who instead of staying with us immediately went to hang laundry; and the new boy, Yoan, who had changed in the meantime and was slinking away, half naked, toward the fields.

I had spent so many hours dreaming of the masseria and of all of them.

In order not to ask about Bern for a third time, I asked where Tommaso was.

"Tommi has his own life now. He works in Massafra, in a resort for rich people. What is it called, Floriana?" He raised his voice so she could hear him.

"The Relais dei Saraceni."

"The Relais dei Saraceni, that's right. Probably whoever came up with the name didn't know what the Saracens did around here." He chuckled and, by reflex, I smiled too.

I could simply have asked him outright: Is it true that Bern got a girl pregnant? But I felt it would have been a slap in the face for Cesare. I saw him lean back in his chair and take a deep breath.

"I don't think we'll be having lunch. Too hot. But you're welcome if you want to stay awhile."

"They're waiting for me at home."

Somewhere Yoan was beating an almond tree to make the fruits drop. You could hear the smacks of the branches he struck, followed by the sound of the nuts falling to the ground, like a hailstorm. Cesare rubbed his face eagerly. "In that case, I'll tell Nicola you're here."

I CAN'T BEGIN to describe the days that followed, the state I fell into. It was something like my fear of the dark when I was a child, when I would stare at the mosquito coil until I could sense the room breathing. There was no reason to stay any longer, except for the remote, irrational hope that Bern might return. Nonetheless, I decided to wait for Nicola's arrival.

I spent many hours in the pool, lying on an inflatable raft. As I lightly pushed myself from one edge to the other, I thought back to the night when the boys had dived in. The pool had been emptied and refilled several times since then, the water treated with chlorine and anti-algae agents, but perhaps some molecules from Bern's skin had survived. I wet my hands and rubbed them over my stomach and shoulders.

My grandmother had continued being as considerate as she was on the first day. She was even willing to leave the sofa and come and read on one of the lounge chairs by the pool to keep me company. She huddled in the patch of shade cast by the umbrella, and once she even put on a bathing suit. Her legs, which I had not seen bare for years, were flaccid and pale, dotted with brownish-colored spots. That afternoon she sat there for a long time, rapt, the closed book in her hands, as if pondering something. Then she turned toward me resolutely and said: "Did you know that your father had almost been married before meeting your mother?"

I clung to the ladder to keep the float from rotating.

"He was your age when he met her. Her name was Mariangela. A beautiful girl."

I slid off the raft, into the shallow water.

"When he told me he wanted to marry her, I nearly had a stroke. I did not agree, but your father is stubborn, as you know. So we made a deal: first he would finish university and then he would marry Mariangela."

I tried to picture the girl, but I couldn't. For a moment, my grandmother turned her head toward the villa. Something seemed to have troubled her.

"So he went to Turin to study at the Polytechnic. When he came back here during vacation he raced over to see her, but as soon as he saw her he realized that she no longer meant anything to him. They broke up that same afternoon. It was a horrible summer for everyone."

She stretched her legs out in front of her and flexed her feet.

"The year after that he met your mother," she added in a neutral tone.

"Does she know?"

"Your mother? Maybe. But I don't think so."

"You think he never told her?"

"Oh, Teresa! It's not a given that because two people get married they tell each other everything."

I had inherited my grandmother's curved toenails. I had not yet made up my mind whether they were a mark of beauty or a flaw. She complained that with age they tended to become ingrown.

"It just goes to show," she added, "that it's foolish to think that the differences between two people disappear only because one wants them to. All your father got for it was wasted years that he could have put to much better use. He and Mariangela would have been happy together, almost certainly."

"How do you mean, happy?"

"Unhappy. I said they would have been unhappy together."

"I thought you said 'happy.'"

My grandmother shook her head. She smoothed her thighs with her hands.

"Look how ugly my knees have become," she said, squeezing them like a couple of oranges.

She turned to smile at me. "There's always a lot to know about someone else's life, Teresa. You never stop learning. And sometimes it would be better not to start at all."

ONE EVENING Nicola came to see me. From the window I saw him standing next to Rosa, the difference in height making her look minuscule. It looked like she was instructing him about something. Nicola nodded, but I couldn't make out their words, and anyway I wasn't interested. I let him wait awhile as I dressed and brushed some mascara on my eyelashes.

Right away I noticed that his manner was different; now he had a kind of studied composure. He suggested we take a walk, but I begged him to go out. My grandmother's villa had started to feel like a prison.

There weren't many people at the Scalo. We sat down at a table in the center of the patio. Nicola went to get us a couple of beers. He seemed proud to finally be able to show me how urbane he was, happy to be alone with me. I immediately regretted having persuaded him to take me there. We didn't seem to be able to get any kind of conversation started.

"Your father says you're doing well at university," I said listlessly.

"He tells everyone that. But I'm really average. Would you like to come to Bari? I could take you one of these days."

"Maybe."

His hands were rather striking, they were big and very smooth on the back. He had overdone the cologne.

"Do you have a girlfriend there?" I asked, to divert him from any fantasy about the two of us together, including a trip to Bari.

He darkened. "Not really."

The strands of lighted bulbs trembled in the wind. Some were burnt out. I wondered if they were the ones from last summer.

"And you?" Nicola said.

"No one special."

But I didn't want him to think I was pathetic. All that time spent waiting for someone I wouldn't see again. I added: "Just a few flings."

"Flings," he repeated, disappointed.

"Him, where is he?"

Nicola leisurely took a sip of beer. "I don't know. He disappeared."

"Disappeared?"

"He left. You probably noticed that he was already acting a bit strange last summer."

"Actually, no."

I didn't know why I was becoming so aggressive, as if it were all his fault.

"Strange how?" I asked.

"He became . . . I don't know. Edgy. Callous, especially toward Cesare."

It always struck me whenever Nicola referred to his parents by their first names.

"Cesare is tolerant," he said. "As far as he's concerned, a person can behave any way he wants, as long as he doesn't offend others. But Bern . . . provoked him. Especially after he started reading all those books and waving them under Cesare's nose."

"What books?"

"Any that stood in opposition to God. Almost every day Bern left one of them for Cesare to find on the table. He highlighted the worst passages so he'd notice."

He'd picked up a dead twig that landed on his bench and was now carving vertical lines onto the pale tabletop.

"There was no reason to treat him like that." He hesitated before continuing: "You know what Cesare told me once?"

"What?"

"That Bern's heart was touched by the evil one."

"The evil one?"

"The devil, Teresa. Cesare knew that the evil spirit lived somewhere in Bern. He prayed every day that it would not awaken, but it did."

"Do you really believe these things?" I asked heatedly.

The twig snapped in his fingers. Nicola looked at it, disappointed, then tossed the two pieces away. "If you knew him well, you'd believe it too."

I did know him well. We'd been together in the reed bed. He had used his tongue on me.

"Just because Cesare says so doesn't mean it's true."

"Bern was angry with him because Tommaso left. Bern said Cesare had driven Tommaso out. But that's not how it was. It's normal for those who turn eighteen to move away from the masseria to go and live on their own. If it hadn't been for Cesare, Tommaso would still be living in that orphanage near the prison. But Bern wouldn't forgive Cesare for it, no matter what. They've always been thick as thieves, those two. Remember the night of Tommaso's birthday, how they cried?"

Instinctively, I turned toward the spot where Bern and Tommaso had lain on the stony ground during the celebration. There was nothing but flat rocks out there. And farther on, the barbed wire, the scrubwood, and the tower. For a moment I thought I could see an animal darting through the nettles.

"And the girl?"

Nicola studied me as if to gauge how much I already knew. If I hadn't mentioned her, I could tell he wouldn't have brought her up at all. He shook his head, as if there was nothing he could say about that.

"Who is she?"

He brought the glass to his mouth, but realized it was empty. He looked a little distraught. Maybe he'd imagined a different kind of evening. I pushed forward the beer I'd hardly touched and he nodded.

"I met her only a few times, because I was always in Bari. She had money issues and . . . I don't know, maybe drug problems too. When she became pregnant, Cesare agreed to take her in at the masseria. She had no place else to stay."

He looked at me. "She had a strange name. Violalibera."

I had the feeling I was falling backward, and I clung to the bench.

"Violalibera," I repeated.

"She's . . ." But he left the sentence unfinished.

"She's what?"

Nicola reached toward my face, his huge hand brushed my hair off my forehead and stroked my cheek with a gentleness I wouldn't have imagined. "I'm very sorry for you," he said.

"I want to go home."

"Right now?"

"Right now, yes."

"Whatever you like."

But more minutes passed before we moved. The Scalo was slow to fill up. The girl who served the drinks was leaning on the windowsill of the trailer looking bored. For a long time we studied each other over Nicola's shoulder, until she widened her eyes as if to ask what the hell I was staring at.

THE FOLLOWING MORNING I told my father that I was going back to Turin. He asked me why, as if he didn't have the slightest idea, and I, as if I believed him, made up a story about wanting to prepare for the start of school. He said that traveling by train alone for all those hours was out of the question, but my grandmother must have convinced him otherwise, because after lunch we went to the station together and bought a ticket for the Intercity departing the next evening.

I packed my bags. From time to time nausea forced me to sit down and breathe deeply. I got angry with Rosa because she'd put a pair of my jeans in the washing machine. Less than an hour later they were pressed and folded on the bed, next to the suitcase.

In the morning I saw her and Cosimo driving away. I don't remember if I thought about it just then or if I had concocted the strange plan during my troubled night, but I took the spare keys to the lodge, went in, and grabbed the jar with the red weevil from the tool shelf. Then I climbed on my bike and pedaled to the masseria at breakneck speed.

I found Cesare kneeling on the ground, fiddling with the septic pit. He wore a pair of high boots and rubber gloves. Yoan stood next to him, leaning on a shovel. A revolting stench came from the trench.

I shoved the jar with the parasite under Cesare's nose and said, "This too? Should we have a funeral for this too?"

He looked at me, dumbfounded.

"Well?" I pressed him. "There must be a soul here too, right? We have to bury it."

He stood up slowly, took off his gloves. "Of course, Teresa," he said softly.

I demanded that they all be called to join in, even Floriana and Nicola. Cesare dug a tiny hole with his forefinger and put the red weevil in it. He read aloud a psalm: "'All our days ebb away because of your wrath, we consume the years like a sigh,'" then Floriana sang without the guitar, and her unguarded voice brought tears to my eyes.

The hole was covered up and I swore that it was over: I would no longer let thoughts of Bern consume me from within.

Afterward I walked through the countryside with Nicola, both of us silent for a long time.

"I'm leaving," I told him. "I don't think I'll come back to Speziale again." I considered whether it would be too cruel to add anything more, then I did it anyway: "I have no reason to come back here."

We skirted a crumbling dry-stone wall. I stopped near a caper flower that had blossomed in the cracks, picked it, turned it between my fingers a few times, then tossed it away.

Topping a rise, we found ourselves unexpectedly in front of the reed bed.

"Why did we come here?" I asked.

Nicola placed a hand on the trunk of an olive tree. He was looking at the ground, not the exact spot where Bern and I used to lie, just to the right of it.

"I asked you why we came here," I repeated. Agitation tightened my throat.

"Bern and Tommaso were brothers to me. They may have been joined at the hip, but I was still their brother."

"So?"

"The three of us shared everything." His eyes bored into mine. "Everything. But Bern never wanted to share you. He said you were his, period."

He ran a hand through his hair. The water of the brook coursed along with a low gurgle. "I have to catch the train," I said. Then I turned and started walking back toward the masseria. Nicola did not move to follow me.

When I was already a distance away, I saw him standing in the same position, facing the reed bed, one arm hanging by his side and the other stretched out to the tree, still spying on the ghosts of Bern and me embracing, or maybe those of Bern and Violalibera, of whoever may have lain on that ground that I, like a fool, had considered mine alone.

From the train I watched the streetlamps streak past the window with its smeared fingermarks, then the long stretches of dark countryside and signs announcing stations of towns I'd never heard of. We must have been in Abruzzo or maybe already in the Marches when a downpour began that in a few seconds fogged up the window and caused the humidity in the compartment to rise oppressively. I had to pee, but I didn't get up, gripped by a kind of paralysis. I had never experienced such invasive pain. The image of Bern and Violalibera together was always there, I kept picturing them over and over again and didn't stop until morning, when a dull sun rose over the plain, finding me awake, still awake.

THE LAST YEAR of high school I studied unstintingly because I didn't know what else to do with myself. It was the only way to prevent my mind from instantly covering the distance that separated me from Speziale. I exchanged a couple of letters with Nicola, but they were dreary and boring, his as much as mine. I stopped answering him.

Even when I was sleeping, the same images played on and on in my head. The boys in the pool. The four of us together in Ostuni, in the center of the piazza, among all those lights. The reed bed and the grueling drives back with my father, he wanting to listen to "Stella Stai" for a second time and I not knowing how to conceal my dejection. In the morning my mother would find me facedown on the desk; she woke me stroking my forehead, then it took hours for my stiff neck to go away.

Every other evening I'd go to the communal swimming pool to wear myself out. The first cigarette I smoked when I came out of there had a funny taste, like burnt plastic; it surprised me every time.

I got the highest grade on the final exams. No one would ever take me for who I really was: a drudge who was trying to forget the guy she'd had an affair with two years earlier, the guy who had gotten another girl pregnant and then disappeared.

In August my father left for Speziale by himself. The morning he left I did not get up to say goodbye. In the days that followed, I didn't call him once.

I was determined not to ask him anything when he returned, but it was he who came into my room, trailing a cloud of perspiration after all those hours of driving. I was watching MTV, a video clip of the Skunk Anansie song "Secretly."

"This year it was hotter than usual," he said.

"So I heard."

"There's a drought so bad, even the old-timers can't remember one like it. It will be good for the olives," he said, sitting on the bed. "But I went to the shore a few times. The sea was perfect."

I turned to the TV. The three characters in the video were turning a motel room inside out.

"Could you turn it off a moment?" he said then.

I looked for the remote control. Instead of turning it off, I muted the volume.

"I was saying that the masseria is completely deserted. There's a 'For Sale' sign."

I asked about Cesare, in a low voice.

"He's gone. I asked around in town as well, but no one knew much about it. They more or less kept to themselves, those people."

He said "those people" strangely, as if he were talking about a group of aliens.

"It won't be easy for him to sell that place. The house would have to be demolished and rebuilt. Actually, I don't know if anyone would even be allowed to rebuild. I'm pretty sure that much of what there is was built without permits. Anyway, who would want that land? Your grandmother says they've been clearing away stones for years."

Finally he stood up, slapping his pants to remove the dust.

"I'd better go take a shower. I forgot, your grandmother sends you this."

He handed me a wrapped package, which felt like a book.

"She was very sorry not to have you there this year."

I tried to picture the masseria deserted, the doors and windows barred, the FOR SALE sign. I watched my father leave.

The images of "Secretly" continued to stream without sound; it was the final scene. I turned off the TV, then unwrapped my grandmother's package. Inside was one of her mystery novels, Martha Grimes's *The Anodyne Necklace*. What drivel, I thought. I put it on a shelf without even opening it.

2.

Many years later Tommaso and I would be the only two left to remember those summers. We were adults by then, past thirty, and I still couldn't say if we considered ourselves friends or the exact opposite. But we had spent a good period of our lives together, the most important time perhaps, and the number of memories we shared made us more alike and closer than we would have been willing to admit.

I hadn't seen him for quite some time, with the sole exception of an evening when I had shown up at his house unannounced and he had chased me away; and I, for spite, had blurted out what had happened to Bern. But on Christmas Eve in 2012 I found myself in his apartment in Taranto, sitting next to the bed he was lying in, so drunk that his arms were trembling a little. He was in such bad shape that he couldn't look after his daughter, Ada, so he'd called me, the last person he'd have liked to ask for help, but the only one he knew would be alone as he was that night.

Around eleven o'clock Ada had fallen asleep on the sofa and I had returned to the room where Tommaso lay. He was awake, staring at the turned-down sheet. Medea, his dog, was dozing, curled up at the foot of the bed. Only one of the bedside lamps was on, the one across from us, and the light would remain on until dawn, until I woke up with the buzz of all the things I hadn't known before.

"The institution," Tommaso began after a long silence, "was a horrible place."

He hissed the words through his teeth, painfully. His skin was grayish, because of all he had drunk.

"Which institution?"

"The one they put me in after my father was arrested."

"What does that have to do with anything now?"

I wasn't there to hear about the institution. Something much more important had been left unfinished between us, something that concerned Bern and Nicola and Cesare, our first summers at the masseria, and Violalibera, that name that from time to time came back into my life.

"For me everything starts there."

"Okay," I said, trying to control my impatience. "Go on."

"There was always a revolting smell, especially in the hall. Soup, piss, or disinfectant, depending on the time of day. So while I sat on the bench, waiting, I breathed in the hollow at my elbow."

His voice became a little clearer with each sentence, as if his lungs and throat were also waking up from his drunken stupor.

"My mother used to say that I'm very sensitive to odors because I'm an albino. She always had the same justification for everything: 'You're albino.' But by then she could no longer say it because she was dead."

He glanced at me, briefly, to see my reaction, but I didn't feel sorry for him. I may have, once, but it had been a long time ago. I just wanted him to continue.

"I sensed them coming even before I saw them, from the smell. Cesare and Floriana, I mean. Soap, mint candy, and the residue of a fart. I was shaking a little, I think. It all seems completely normal now: you're ten years old and you're waiting for strangers to take you away. Floriana sat down and stroked my hand without taking it, while Cesare remained standing. I didn't take my nose out of my elbow and didn't look at them directly; all I saw was his shadow that stretched across the floor to the wall. He touched my chin, forcing me to raise my head. He still had a beard, and when he was upset he combed it with his fingers. He did so after telling me his name. But I already knew it, the social workers had told me about Floriana and Cesare, and had shown me a photograph in which they were hugging each other in front of a yellow wall. 'Two devoted people,' one of the caseworkers had said.

"'Look at him,' Cesare said to Floriana, 'doesn't he remind you of the archangel Michael? In that painting by Guido Reni.' Then he whispered to me: 'The archangel Michael defeated a terrible dragon. I'd like to tell you his whole story, Tommaso. But we'll have time in the car. Gather your things now.'

"In the car, however, he did not continue the story. He only added that their house was on the archangel's line, a line that went from Jerusalem to Mont Saint-Michel. Maybe that was the whole story.

"I tried to memorize the way, the direction from where my father was, but I soon lost my bearings.

"'I'll take care of the bags,' Cesare said when we got out of the car. 'You go look for your brothers.'

"'I have no brothers, sir.'

"'You're right. I spoke hastily, I'm sorry. You will decide what to call them. But hurry up now, they must be around here. Past the oleanders.'

"I went through the shrubs and wandered awhile in the olive grove, at first not venturing too far from the house. I still had the foolish hope of being able to run away, of finding the institution not far off. I wasn't used to being in the country. I was about to turn back when I heard a voice calling to me: 'Over here!'

"I whirled around, but I didn't see anyone, only the trees set apart from one another.

"'In the mulberry tree,' the voice said.

"'I don't know which one is the mulberry tree.'

"There was a silence, then I heard footsteps. Bern popped out of the shadows.

"'This is the mulberry tree,' he said. 'See?'

"I went over; it was dark and cool under there. A ladder led to a small treehouse built up among the branches. Bern studied me. He touched my cheek and then he said: 'You're so white, you look very delicate.' I replied that I wasn't at all delicate. He climbed up the ladder and I followed him.

"Inside the little house, sitting cross-legged, was Nicola.

"'See how he looks?' Bern asked him, but he barely glanced at me.

"'At least this one is brave enough to come up.'

"As a matter of fact, the treehouse didn't give the impression of being very stable. I asked if they had built it themselves, but they ignored me.

"'Can you play skat?' Nicola asked.

"'I can play poker.'

"'What's that? Sit down, we'll teach you skat. We were missing a third.'

"They listed the rules in confusion, talking over one another. For the rest of the afternoon we didn't say another word except for those related to the game. Then they said it was time for prayer. I prayed at the institution too, so it didn't surprise me. I couldn't imagine how different it would be. We went down the ladder one at a time. Once past the oleanders, the naked bulb shining under the pergola appeared. Bern put an arm around my neck, and I let him do it. I had never had brothers, and before that day I didn't even know how much I wanted them."

TOMMASO PAUSED. It seemed to me that a calmness had spread through his body as he remembered his first day at the masseria. I knew that feeling, the treacherous consolation of every memory that had to do with that place.

"Cesare's eyes illuminated everything," Tommaso went on, "the masseria and the surrounding land, but especially us boys. All you had to do was let a sigh escape you during lessons and he would grab you by the arm and say, 'Come with me, let's talk a little.'

"Under the holly oak he was prepared to wait as long as half an hour to get a word or a sign out of you. I didn't know what Cesare wanted from me. But he waited some more, and even longer, his eyelids half closed, perhaps dozing. Then, suddenly, a word would pop out of my mouth like a saliva bubble. Cesare would nod to encourage me. Another word would follow and eventually everything came pouring out. Then it was his turn. He commented at length, as if he'd known from the beginning what I would confess to him. We'd pray together to beg for mercy and wisdom, then we'd rejoin the others, and for a few hours I felt light and clean.

"The only place we were safe was in the little house in the mulberry tree. The foliage was so dense that Cesare couldn't see us when we were up there. He would come to the base of the trunk and call up from below: 'Everything okay?' He'd peer through the planks, but we had covered the bottom with a tarp we'd found in the tool shed. Then he'd lose his patience and walk away. Sometimes I think I was the one who brought perversion to the treehouse. For sure I was the one who taught Bern and Nicola the swear words I'd heard in the institution's mess hall. They took turns repeating them, to savor their tang. A video game had passed the inspection of my bags unscathed and they used it avidly as long as the batteries lasted. I remember that for a time we dared one another to taste all the leaves, roots, berries, seeds, and flowers on the masseria. The thought of who would be first to stumble upon something poisonous excited us.

"One afternoon Bern found a wounded hare next to the woodshed and we took it to the treehouse. It peered at us with glassy eyes that were bright and anguished. A yearning came over all of us. 'Let's kill it!' Bern said.

"'We'll be damned,' Nicola replied.

"'No. Not if it's a sacrifice to the Lord. Tommi, hold it up.'

"I grabbed the hare by the ears. I could feel the rapid pulsing of its heart in my fingers, or maybe it was my own pulse. Bern opened a pair of scissors and passed a blade across the animal's neck. But he was too gentle, he didn't succeed in cutting it. The hare twitched, and almost got away from me.

"'Cut!' Nicola shouted. Now his eyes were feverish.

"Bern pulled the hare downward by its good paw; stretched out like that, it was very long. He closed the shears and plunged them into the animal's throat like a dagger. I saw the point press into the fur on the opposite side, without piercing it. When he pulled the scissors out with a spurt of dark blood, the hare was still flailing.

"Bern stood frozen, the scissors in his fist. Now it seemed as if the hare were begging him to use them, to finish the job as quickly as possible. Nicola elbowed him aside, shoved the shears into the open wound, and promptly snapped them open. Blood splattered across my face.

"We buried it as far away from the house as possible. Bern and I dug with our hands while Nicola acted as lookout. When we returned to the burial site, a few hours later, there was a cross stuck in the ground. Cesare said nothing about it, but that evening he read a passage from Leviticus with long, eloquent pauses: 'The hare, which indeed chews the cud, but does not have cloven hoofs and is therefore unclean for you. The pig, which does indeed have cloven hoofs, but does not chew the cud and is therefore unclean for you. You shall not eat their meat, and you shall not touch their carcasses; they are unclean for you.'"

"THEY ARE UNCLEAN for you," Tommaso repeated. Then he said it again, whispering: "Unclean."

He cupped his hands loosely, rapt in something.

"But it was Nicola who got us the magazines," he continued after a moment. "Sometimes Floriana sent him to town to run some errands. She had countless ways of favoring Nicola over us. To me, it didn't matter; once a month I too was permitted to go into the city to see my father. But Bern couldn't tolerate it. When Nicola or I came back from our brief outings, his eyes would rake us over, even if he said, 'Out there? So tell me, what's out there that should interest me so much?'

"Nicola had spotted the magazines on the display rack at the newsstand. Just a glance, at least so he'd thought, but the news dealer had invited him to take a couple: 'I'm giving them to you. Don't worry, I won't tell your father.'

"We had a long discussion in the treehouse before opening them. We decided to look at only two pages a day, the sin would be less that way. We three talked a lot about guilt, the Commandments, and sin. In any case, we did not respect the agreement. We went through the whole magazines that same afternoon, shocked and unsmiling, as if we were looking straight into the infernal abyss. Even then I knew that I was focusing on the wrong details, I was scrutinizing those photographs differently than my brothers, but they didn't notice it.

"Bern and Nicola lowered their shorts. It was the beginning of June, the clusters of mulberries left purple smears on the wooden planks, our elbows and knees.

"'You too,' Bern said to me.

"'I don't feel like it.'

"'You too,' he repeated. And I obeyed him.

"We forgot about the magazines. We didn't need them anymore and we wouldn't own any others. All we had to do was look at one another. In the evening, at supper, our shame was so great that it made us completely unreadable to Cesare.

"Other boys lived with us over time, but we kept them at a distance, we didn't allow anyone to climb up into the tree. They only stayed for short periods anyhow, and one morning, without any notice, they were gone.

"Finally, the treehouse in the mulberry became too small. The last one to climb up there was Nicola. He found a hornets' nest lodged among the branches. We always said that we would build a new, more spacious refuge, maybe over several trees connected by rope bridges, but time had begun moving faster than us."

HE PAUSED and began counting silently on his fingers, very slowly. A part of me still wanted to urge him on, but another wanted to lose myself in his recollection of the early years at the masseria, to relive the warmth that I too had known there.

"It was still '96," he said, "September 1996. Nicola entered the last year of high school in Brindisi, to study humanities. To catch up, he had taken private lessons from a teacher in Pezze di Greco. He had moved out of the room we shared to the one where Cesare kept his oil paintings, to concentrate better. That room was always kept locked. Naturally, we'd broken in many times and we knew that the subject of the paintings was always the same: a meadow dotted with red flowers in which olive trees grew, with one flower much taller than the others in the foreground. That giant poppy was him, Cesare, that's obvious, isn't it? But I don't know if I understood it so clearly back then.

"Nicola now had brand-new textbooks, an English dictionary of his own and a Latin one as well, while Bern and I continued to consult the one that was split in three from wear and tear, the words practically illegible. Nicola forbade us to touch his books, he said they were expensive. In the morning he would leave in the Ford with Floriana and return after lunch by bus. He was exempted from chores on the farm because in the afternoon he had to study, so his tasks were distributed between Bern and me and ate into part of our lessons. In any case, Cesare didn't seem to want to spend as much time with us.

"And then the computer arrived. Two big boxes on the kitchen table. The technician opened them with a cutter and took out the parts protected by polystyrene. After all the years spent with Cesare, I wasn't used to technology, we didn't even have a radio. And now, all of a sudden, a computer! In our house!

"'In my room,' Nicola explained to the technician, who was pointing to an outlet in the wall.

"Bern leaped up. 'Why?'

"He blocked the technician's way, and almost made him stumble.

"When he saw that he couldn't stop him, Bern asked, 'Can we use it?'

"Cesare had put on his reading glasses to make out the small print on the box, but it was incomprehensible to him, you could tell by the worried arch of his eyebrows.

"'Can we use it or not?'

"Cesare took a deep breath. He spoke looking directly at Bern, undeterred, yet, perhaps for the first time since I'd known him, his voice seemed to show some uncertainty: 'The computer belongs to Nicola. His teacher . . .'" He stopped. 'Be patient. Your time will come, too.'

"Floriana was leaning on the kitchen counter, looking tight-lipped at her husband, and I could tell it was a decision that they had made together.

"Meanwhile, Bern was on the verge of tears, the computer now placed in the only room of the house that was off-limits, the object of an irrepressible desire that a moment before he didn't know he had.

"'Based on what principle?' he asked.

"No one answered. The technician unwound the cables and connected them.

"'Based on what principle, Cesare?' Bern repeated.

"It's at that moment that something ruptured between them, in the pause between the question and the answer. Cesare said: 'You shall not covet your neighbor's wife, nor his field, nor his slave, nor . . . ,' but he was interrupted by the slamming of the door.

"Later, in the bedroom, Bern vented to me: 'It's not fair. They already gave him the room.'

"'Nicola is older,' I said.

"'By a year. It doesn't even bother you that he goes to school, while we're stuck here,' he accused me. 'You don't want to learn anything. And you don't care about anything.'

"But it wasn't true. Talking with him in the darkness, or listening in silence to the drops that fell from the eaves after the evening rainstorm: that was what I cared about, and it was better than anything I had ever had. Why couldn't he be satisfied as well?

"'How do you think they pay for Nicola's private lessons?' he persisted.

"'I don't know. With what Floriana earns?'

"Something hit me in the face: a balled-up sock. I threw it back at him.

"'And how do you think they paid for the computer, you fool? Cesare gets paid to keep you here.'

"I didn't want Bern to talk about it. The indemnity compensation, that's what they called it on the foster-care forms. I could feel the indemnity compensation pinned to me, like the price tag on a new T-shirt.

"'So?' I said.

"'So my mother also sends money to Cesare, what do you think? Even if he's her brother. She sends it to him every month. And he benefits from it.'

"I saw his shadow sit up in bed.

"'Starting tomorrow we're on strike,' he announced.

"'What does that mean?'

"'We'll do what the baron did the day he escaped up into the trees.'

"'Oh, sure. Cesare would make us come down immediately.'

"And yet I would have done so if Bern had asked me, I was ready to do anything for him, even if it meant never setting foot on the ground ever again.

"'Right, we'll do what the baron in the trees did,' he went on, as if talking to himself by then. 'It's his approach we have to follow. From tomorrow on, no more lessons. No more prayers. No more chores.'

"I turned around to face the wall. One night, years ago, we had slept in the treehouse with the excuse of seeing shooting stars. Toward morning the air was so cold and damp that we had returned home, barefoot. Cesare had brought us cups of steaming chamomile tea to warm us up. He'd been good to me, more than anyone else. He didn't deserve my disobedience.

"'So are you with me?' Bern asked.

"The next day we gathered under the holly oak to sing God's praises. Cesare only allowed us to wear our tunics in the morning. When we first awoke, he said, we were purer.

"He read something from Ezekiel, I heard only a word or two. He's over it, I kept telling myself with relief, sleep has calmed him down.

"Cesare asked him to find the passage about the Garden of Gethsemane in Matthew's Gospel. Floriana handed him the Bible and Bern opened it. He was faster than us at finding the verses, almost faster than Cesare by now. He held the open book in front of him, took a breath to begin reading, but no sound came out of his mouth.

"'Go on,' Cesare encouraged him.

"Bern glanced briefly at the sky, then back at the book.

"He closed it again. 'I won't read,' he said.

"'You won't read? Why not?'

"Bern's cheeks were flaming red. I was hoping he wouldn't bring up the computer right then, if he did it would seem ridiculous, even to me.

"But Cesare understood just the same. He uncrossed his legs and reached over to take the Bible back from him. He gave it to me.

"'Tommaso, you read for us this morning, if you would be so kind.'

"The olive trees embraced us all around. We could indeed have been the disciples gathered in Gethsemane.

"'Luke?' I asked, leafing slowly through the pages.

"'We'd said Matthew,' Cesare corrected me, '26:36.'

"I found the place. Bern was waiting for my show of loyalty. But he'd forgive me no matter what. Sure, he'd get over it sooner or later. My calves, crushed under my buttocks, had pins and needles.

"Instead Bern cried out: 'Don't read!'

"There was no arrogance in the way he said it, if anything it was a supplication.

"'We're listening, Tommaso,' Cesare urged me.

"'Then Jesus came with them to a place called Gethsemane . . .'

"'Don't read, Tommi,' Bern said more softly. He knew he had me in his power now.

"I put the book down. Cesare took it and gave it to Nicola with a patient gesture. Nicola began to read, stumbling continually, the phrases disjointed due to his discomfort. He hadn't yet finished when Bern jumped up. Crossing his hands behind his head, he rolled up the tunic and pulled it off. He tossed it on the ground like a rag, and stood there in his underwear. He was breathing rapidly, I could tell by the jerking of his shoulders. He seemed so vulnerable, so angry.

"The only sound was that of the leaves stirred by the wind. I bent forward to remove my tunic as well, but I was more awkward than him. Cesare, however, had stopped looking at us. With his eyes closed, he intoned the 'Alleluia.' At the second stanza Nicola and Floriana joined him, their eyelids lowered, as if to refuse to look at that naked, infidel version of us. Bern broke the circle and strode off toward the house. I followed him, driven by the accusatory singing of Nicola and his parents. When I was halfway there I turned to look at them sitting under the tree. I stood like that for a few seconds, suspended between the three of them and Bern, two families that were suddenly split, neither of which, I realized instantly, would ever really be mine.

"The strike lasted until the beginning of the summer. The first week, Cesare did not abandon the hope that it was nothing more than a passing whim. He would sit under the pergola with the books neatly stacked, and from there he kept giving us looks that made me sick to my stomach. But after a while he got tired and stopped waiting for us.

"He developed a strange cough. One day he had a long, vigorous outburst, and behind Bern's back I brought him a glass of water. He accepted it, then took my hand and pressed it to his chest.

"'Love is imperfect, Tommaso,' he said. 'You understand that, don't you? Every human being is imperfect. If only you could make him see reason.'

"I drew my hand back and left him to himself. After that, he didn't ask for my help again, he stopped bothering with us altogether. He allowed Bern and me to sit at the table, he still poured water into our glasses and tinted it with a drop of red wine, but it was as if we were strangers. We didn't talk anymore, we didn't sing.

"One evening Nicola lost control and went at Bern, slapping him. Instead of reacting, Bern slowly turned his head and offered him the other cheek so he could hit him again. His smile was mocking. Cesare stopped Nicola's arm, then made him apologize. Floriana left the kitchen, leaving her plate half full; I didn't remember her ever doing that before.

"'For how much longer?' I asked Bern when we went to bed.

"'For as long as it takes.'

"Together we had not yet stopped praying, but as the weeks went by, new desires emerged in Bern. More than once, opening my eyes during the night, I saw him standing in front of the window, listening to the sounds of distant festivities, watching silent fireworks on the horizon. He wanted to be there, whatever was going on.

"'Don't worry,' he'd say without turning around, 'I'll take care of you.'"

TOMMASO DRANK a little water. Swallowing seemed painful for him, all that talk must have left his throat parched.

"And then the palms began to die," he said. "Word spread among the farmers that the parasite would also attack the olive trees and that the palms should be eliminated as a preventive measure. There was one palm tree at the masseria. For days Cesare wondered what to do about it. He circled around it, studied it. In July there was a scorching heat wave. I don't know if Cesare was afraid the wind would carry the parasites from the south. But one morning we heard the buzzing of the chainsaw. From under the pergola we saw him on a ladder propped against the palm tree. One by one, the fronds dropped. When he was done with those, he got busy on the trunk. The blade slid against the bark, and I squeezed my eyes shut a few times, thinking it might escape his grip.

"With his fists on the table, Bern said, 'He won't make it.'

"But Cesare managed to cut into the trunk, and from there it didn't take much to widen the wedge. The top of the palm remained upright for a moment. Then it leaned to the side opposite the slash and crashed to the ground.

"Cesare ran a rope under the trunk. Tying it around his waist, he started dragging the palm tree's carcass. He was looking for a clear area to burn it. The trunk trailed a few yards behind him, but he groaned and fell to his knees, drained.

"'We have to help him,' I said. My heart was beating out of control. I was afraid he might have broken his back while we looked on uncaringly from the pergola. I took a step toward him, but Bern held me back by the arm: 'Not yet.'

"Cesare got back on his feet, shifted the knotted rope from his pelvis to his shoulders, and, like an ox, began hauling again. The trunk jolted along, but he fell again, gripped by a fierce coughing fit.

"'He'll hurt himself!'

"Then Bern seemed to suddenly come to. We walked over to Cesare. Bern offered him a hand to help him up, then gently stroked his sweaty forehead.

"'You'll let us go to school like Nicola,' he said.

"'What do you hope you'll find there, Bern?'

"Cesare's voice was shaky, and not just from the exertion and the harsh coughing.

"'You'll let us go to school,' Bern repeated, gently fingering Cesare's torso where the rope had left a red burn mark.

"'I've been praying so much for you. Night and day. So that Our Lord would illuminate your hearts again. Do you remember Ecclesiastes, Bern? Whoever increases knowledge increases grief.'

"Bern went on wiping away the sweat from his neck, from his chest, with a tenderness that I would have liked him to use on me.

"'Will you do it?'

"Cesare bit his lips, which were cracked by the wind. 'If that's what you want,' he murmured.

"But Bern hadn't finished, not yet. 'You'll let us go out with Nicola,' he said, 'even in the evening, whenever we want. And you'll give us some of the money you get to keep us.'

"Something appeared in Cesare's eyes. 'Is that what it's about? Money?'

"'Will you do it?' Bern insisted, already tying the rope around himself.

"'I will.'

"'Tommi, go to the tool shed, see if there's another rope.'

"As I rummaged through the tools, I wondered whether Bern knew it or if I was the only one aware of it: whether he knew how much Cesare loved him above anyone else, above his own son—though he could not admit it to Floriana or to himself and maybe not even to God. Because, though it was true that Bern shared only a fraction of his blood, their souls were identical.

"After that day, there was a truce. A certain normality was restored, but nothing was the same as before. Now, during prayers, we held hands with a kind of reserve. Floriana had become openly sullen. Looking back, I'm sure she'd suggested that Cesare send us both away, and that he refused. One afternoon, while we were picking tomatoes, I saw her staring at one that was overly ripe, before angrily crushing it to a pulp in her closed fist.

"Bern and I tore down what was left of the treehouse. By now, any promise that existed for us lay beyond the iron bar across the dirt track.

"The first afternoon the three of us—Nicola, Bern, and I—ventured out alone in the car, we drove south, down to Leuca, to see how far away we could go. We walked around the lighthouse and Bern thought he could make out the profile of Albania. On the way back we got lost in the maze of streets around Maglie.

"In the evening we'd go looking for good times. Nothing ever happened in Speziale, and in any case we weren't well liked there. Once, music led us to Borgo Ajeni, where there was a festival. Smoke laden with animal fat billowed from the stalls. Bern and Nicola held their noses; they would have left immediately if they hadn't been attracted by the crowd and the band that was playing. The smell of roasting meat made me ravenous. I ate it every time I saw my father, but neither of them knew it.

"Bern must have seen something in my eyes. 'I'm getting some,' he said.

"'Don't!' Nicola tried to dissuade him.

"But Bern was already leaning toward the woman who was flipping the ground meat on the grill.

"I ate a sandwich and stopped, but he wanted another and then another, like a drug addict. The grease glistened on his lips and chin.

"Nicola darkened and ended up not having any fun at all. 'You're murderers,' he said as we returned to the car."

FOR A WHILE Tommaso focused on touching his fingertips together, pinkie to pinkie, ring finger to ring finger, and so on, as if to gauge his level of lucidity. "Then we found the Scalo," he said in a neutral tone.

He snapped his fingers and Medea got up right away, reached out her snout to sniff his hand and then lick it. Tommaso wiped it absently on the blanket.

I gave him that moment to catch his breath, and in the meantime I tried to picture them, him, Bern, and Nicola, drawn here and there by music in the air on those summer nights, like strays, then arriving at the Scalo for the first time.

But when Tommaso started talking again, he didn't pick up from that evening.

"Freedom allowed Bern to go to the municipal library in Ostuni and borrow all the books he wanted. After lunch he'd retreat behind the house and read with his back against the wall. During those hours I would sneak into Nicola's room. The computer hadn't been purchased for playing video games, but a few had come installed with the operating system and we got others thanks to some new acquaintances at the Scalo. We played one at a time, without a joystick, using the cursors quietly, to prevent Cesare and Floriana from hearing us. There was a level of Prince of Persia that we couldn't get past. We'd get killed in turn. One day, while I was waiting for Nicola to fail, a movement outside the window caught my attention. It was then that I saw you."

Tommaso glanced over at me.

"You saw us?"

I understood what he was referring to, not the exact day, not that, but I knew he was talking about the afternoons Bern and I were together, our secret hours.

"You were crossing the little clearing that separates the house from the oleanders," he said. "I have a very sharp image: Bern's tanned back with its jutting shoulder blades, you slightly sharper, in an orange sundress. Nicola didn't notice, absorbed as he was by the video game. I was about to say 'Look,' but something stopped me. I watched you disappear behind the bushes. There was nothing in that direction, except olive trees and hiding places.

"'Tommi, you're up,' Nicola told me.

"'What?'

"'It's your turn. Destroy that fucking skeleton!'

"'You go on. I don't feel like it anymore.'

"I went back to my room. I lay down on the bed, but when I closed my eyes I could see Bern walking on the reddish soil with you. So I jumped up, ran down the stairs, and went out. I glanced up at the window, Nicola hadn't moved. A lizard darted in front of me, then scampered up a trunk. I got as far as the mulberry tree, sure you were headed there, and for some reason I felt relieved that it wasn't so. They must have gone to the blackberry bushes, I told myself then. I went from the shade of one tree to that of the next one, wanting to protect my shoulders from the sun. By then I was certain I'd lost you, when I saw a figure standing in the reeds. I moved closer still. It was Cesare. He was watching something among the stalks and it looked to me as if his stocky torso was shaken by a tremor. He wore only his shorts and sandals, he must have left his room like that. I was about to call him, but he suddenly turned and started running in my direction, rustling the reeds.

"He was running toward me, Cesare was, and it was strange to see him running because he never did. When he saw me there he was distraught. For a moment we faced each other, barely a fraction of a second. His arousal was mercilessly revealed by the bright midday sun. He covered himself with a hand and then darted to my right.

"Yet I still hadn't understood what was concealed behind the emerald-green barrier of reeds. Until I saw you two emerge from that little wood, still cautious as you'd seemed from Nicola's window, but with a different look, more disarrayed and exhausted, more complicit, as if you had just swum out to sea together. I hid behind the trunk of an olive tree before you could see me."

TOMMASO'S VOICE had grown fainter, and when he stopped talking, it seemed as if the silence had absorbed it completely. I wanted him to move on, to leave Bern and me in peace in the reed bed. He had no right to intrude in those memories.

He cleared his throat.

"In the hours that followed, Cesare and I avoided running into one another, and when we did we looked the other way. I felt betrayed all around: you and Bern, Cesare lurking in the reeds, Nicola looking ahead toward his new life in Bari. At supper Cesare launched into a longer prayer than usual. He held Floriana's hand and squeezed his eyelids shut so tightly that when he reopened them he had white streaks on his temples. He rummaged for something in his pants pocket and pulled out a folded sheet of paper.

"'I wanted to read you this homily. I remembered it today, after a long time.'

"Did he glance at me before continuing? Maybe, I don't know for sure anymore.

"He read: 'Not even the Father is impassible. When we beseech Him He has pity and compassion, He knows something about the suffering of love, He has failings that His sovereign majesty would seem to proscribe.' For a few seconds he remained there undecided, standing among us as we sat.

"'These failings that we all have,' he added then. 'All of us. Sometimes we are not able to fight them. We would like to imitate Jesus, but . . .'

"He paused again. He seemed more and more confused.

"'It's late. Let's eat now.'

"I knew that he had chosen the homily for me. Was he excusing himself? Was he asking me to forgive him? He couldn't imagine how much I stood by him. Maybe the others loved him because they thought he was infallible, but not me. I loved him, period.

"That night, at the Scalo, I hid behind the trailer and drank as much as I could. I have no recollection of our return home, but I remember that once we were in our room, Bern came over to the bed and put a hand on my forehead. He asked me if I wanted some lemon juice and I told him to leave me alone. In the morning Cesare motioned for me to join him under the holly oak. He was sitting on the bench with the expression he wore on the best days. He had put on his tunic. He tapped the vacant spot next to him.

"'I got up very early,' he said, 'it was still dark, I think you boys had just gotten home. I went into Nicola's room and then into yours; I hadn't done that for some time. I watched you for a while as you slept. It is always miraculous to gaze at innocence while it's sleeping. And all of you are still the mirror of that innocence, even if you no longer think so. You still are, yes, even if you've grown beards on your cheeks.'

"Which wasn't true. My facial fuzz could only be seen in the right light, like that of girls.

"'I was thinking about when Floriana and I came to take you home. I remember telling her: this boy is destined for a special future,' Cesare went on, his fingers smoothing the lower part of his tunic and tucking the edge between his knees. Under the holly oak we boys weren't supposed to talk unless we were questioned, so I kept quiet. 'It seems like yesterday, and yet it's been . . . how many years?'

"'Eight.'

"'Eight, good Lord! And in a few days you will turn eighteen, a man to all intents and purposes. But it seems to me we've already talked about this.'

"'I think so.'

"'So, Tommaso, as you know, the time has come to make your way, to follow your path.'

"I felt my body sag. 'I thought I would stay until I finished school. Until graduation, I mean.'

"Cesare put his arm around my shoulders. 'Oh, that certainly would have been a possibility. If I had continued to be your teacher, that of all of you, of course. But public education is calling you boys, isn't it? Don't worry, I understand that desire. Our Lord understands it too, perhaps he allowed it to germinate because he has a specific plan for you. And who are we to oppose Him? When I think about it, at your age I was planning my first trip. I didn't have a penny in my pocket, but I made it to the Caucasus by hitchhiking.'

"There was really no way to be comfortable while sitting on the bench; I think that was one of the reasons why he brought us there. But he would say: It's impatience that always makes you boys squirm and fidget.

"'Now that you're going to a real school, as an adult, there's no longer any reason to stay here. I spoke to an acquaintance of mine, his name is Nacci. He owns a resort in Massafra. A magnificent place, perhaps a little opulent for my taste, but enchanting.'

"'Massafra is more than an hour away by bus.'

"'But there's a school there, too, what do you think?' Cesare replied with a smile. Then he suddenly turned serious again. I thought I saw the expression he'd had the day before, that fraction of a second when he'd stood frozen before me.

"'We've already agreed. You can move there next week. They'll welcome you warmly and Nacci has sworn that the work will not be hard. You'll earn some money by day, and in the evening you can attend classes in the city.'

"'Does Floriana know?' I asked. Maybe she would persuade him to keep me there for a little while longer.

"'Oh, she was the one who thought of Massafra! It hadn't occurred to me.'

"'And the others?' I asked in a whisper.

"'We'll tell them later. Together, if you like. Give me your hand now.'

"I opened my limp palm, and he took it and squeezed it. I wondered if that was really the last time I would touch those perspiring fingers. I had the right words ready in my throat, 'I won't say anything about what I saw, I promise!' but those were not the kind of words Cesare permitted under the holly oak.

"'Let us pray for this new chapter of yours,' he said. 'May the Lord be at your side at all times.'

"But I didn't listen to his prayer. I was looking toward the house: Nicola on the swing-chair, Bern pushing the seat to needle him, the clusters of tomatoes and onions hanging on the wall. A hoe abandoned on the ground. I couldn't believe that my life was ending like that, all of a sudden, yet another time."

———

"SO THAT'S HOW you started working at the Relais," I said.

But Tommaso ignored me. His expression had hardened. "It was Floriana who drove me there. When I saw the Tropical Pool, with the little bridges and the huge fountain in the middle, I couldn't believe it. It was all so flamboyant."

He took a deep breath.

"On the first day Nacci wanted to know what had gone wrong in my family to make me end up under Cesare's guardianship. I told him in a few brief words, and when I finished he said: 'Jesus Christ! Why would a man do something like that to his wife?' Hearing him talk like that made me realize how far I'd come from the masseria. Nobody there would have sworn. Not that Nacci ever treated me badly, but he was different than Cesare, I could tell right away. He never seemed like a father. After enrolling in the evening courses, I never even attended one class and he didn't urge me to go, maybe he didn't even notice. I started out calling him 'Mr. Nacci' and continued doing so until . . ."

Until the night of the incident, I told myself. I was sure that Tommaso had been interrupted by the same thought. Though maybe he didn't call it "the night of the incident" in his mind. Who knows what words he used in his own head.

"I lived in the employee quarters," he went on. "During the summer season, because of the constant receptions, and during the autumn months because of the harvest, there would be as many as seven or eight of us sleeping in the same room, in bunk beds. There were no screens on the windows to keep the mosquitoes out, so the nights were punctuated by slaps. Smacking my arms or neck, I thought of the masseria, where it was forbidden to kill even the tiniest creatures. When I heard the buzzing in my ear again, I felt almost relieved. Nacci realized that I wasn't skilled in anything, that I had never cleaned a swimming pool, didn't know how to serve a table, and knew only a little something about plants, so he introduced me to Corinne. He told me to remain glued to her for as long as it took. 'For as long as necessary,' he said, but I don't know if he really meant that long."

Tommaso smiled. Then he pulled up the edge of the sheet to cover himself better. To protect himself from his own joke, I thought.

"The first thing Corinne said to me was, 'You look like the crazy replicant in *Blade Runner*.' She didn't say it to be funny, far from it, she was dead serious. When she was a few steps away, Nacci whispered in my ear, 'Don't listen to everything she says. And don't trust her too much. She's a junkie.'

"Corinne taught me how to keep my shoulders back as I made my way around the tables and how to bow slightly when I offered the guests a tray. During role-playing sessions she was always the guest, a capricious guest who used any excuse to humiliate me. 'You have to get used to it, Blade. They take it for granted that they're better than you just because you're serving them drinks.'

"She showed me how to open the wine and sniff the cork, how to pour water, and when I proved that I could do everything even better than she did, she declared the lessons were over.

"In October, I wore the uniform for the first time. At the Relais they were celebrating the wedding of an actress, I didn't know her, but she seemed lost in all the confusion. When I saw that she hadn't eaten anything, I brought her a little plate of sliced fruit on the sly. 'Eat this at least, or you'll feel dizzy,' I said.

"In the kitchen Corinne came up and surprised me from behind. 'Why did you do that?'

"'Why did I do what?'

"'*Eat this at least, or you'll feel dizzy*,' she mimicked me, whining.

"'Did I do wrong?'

"She rolled her eyes. 'You're really a drag, Blade! You take everything seriously,' she said, punching me in the stomach, hard. I suspected it was an excuse to touch me.

"Later, in the locker room, she sat down on the bench and stayed there staring at me as I took off my jacket and shirt and finally my pants.

"'Want to see a place?' she asked.

"I looked at the clock on the wall.

"'What's the matter, too tired? No problem.' She stood up to leave. Always that aggressive tone. Even then I couldn't stand up to her.

"'Okay,' I said.

"I followed her through the dim rooms, to the door that led to the vaults. 'I've already been down in the cellar. Anyway, it's locked.'

"Corinne fumbled in her jeans pocket, pulled out a key.

"'Ta-dah!'

"'How did you get it?'

"'A guy who worked here,' she said, turning the lock and opening the door a crack, all without the slightest sound. 'If you tell Nacci, I'll slit your throat, Blade, I swear.'

"We walked through the equipment and steel tanks.

"'Sit down there,' she ordered.

"'On the ground?'

"'Are you finicky?'

"I watched her grope around for something behind one of the tanks. She found a glass and filled it from the tap at the bottom. She drained it in one swig, then filled it again. 'Doesn't serving wine to all those guests make you want some?' she asked me.

"I drank, and felt her watching me in the dark. When I handed back the glass she said, 'I thought it was forbidden for you Jehovah's Witnesses.'

"'Who said I'm a Jehovah's Witness?'

"'That's what they say.'

"'I'm not a Jehovah's Witness.'

"'You can be whatever you want, I don't care,' she said, turning her head to look me in the eye.

"'Anyway, they say things about you too,' I hit back.

"Then she leaned her face over, very close to mine, as if she wanted to bite my nose, and I didn't dare back away or turn my head. Whispering, she said, 'Do I scare you, Blade?'

"I sat stock-still until she moved back, sneering. 'Let them say what they want. They're just curious. And envious. Come on, ask me a question. What do you want to know? If I used needles? If I exchanged them with other people?'

"'I'm not interested.'

"'It's what everyone is interested in. Whether I exchanged needles and where I got the money. People have wild imaginations. Do you have wild fantasies too, Blade?'

"'No.'

"I didn't dare look at her anymore, so her hand on my neck caught me by surprise. Just a caress, light.

"'If only they were all like you,' she said.

"Then she got up to draw more. We passed the glass back and forth for a while.

"'Have you been to Jakarta?' I asked her finally.

"Corinne sank her chin into her sweatshirt. 'No. My father gives them to me. I once told him that I liked the Hard Rock Cafe and from that time on he brings me one from every city. I have five continents so far.'

"'He's a diplomat,' she added with a trace of irritability. 'Before the age of thirteen, I lived in . . . Let's see if I can remember,' she said, and began counting on her fingers. 'Russia, Kenya, Denmark. And India. But only for a few months.'

"I pictured each of those countries in the color in which it appeared on the tablecloth at the masseria. I could see the entire tablecloth distinctly, as if it were spread out before my eyes.

"Corinne had started stroking the raised yellow circle on the sweatshirt. 'I told him three years ago that I didn't like them, but he keeps bringing them to me. I wear them to go to the gym.' She finished what was left and got up yet another time. But when she turned on the tap she hesitated. 'I'll show you a way to get high faster,' she said. 'Get up. Come on, stand up! Climb up here.'

"I did as she said. I climbed up the ladder on the side of the tank. When I got to the top, Corinne explained how to open the top hatch. 'Keep your head back. If it gets into your eyes, the first fumes will blind you. Inhale slowly.'

"I inhaled the vapor and it was like a wallop, I was nearly sent flying backward. It was as shocking as the time Nicola and I had gotten our hands on Floriana's stash of distilled spirits, but the effect of the liqueurs was nothing compared to this. I leaned over and breathed in again, after which I don't know how I managed to get down. But I know that I burst out laughing, I laughed so hard that Corinne had to cover my mouth with her hand. Since that wasn't enough to stop me, she wrapped her arms around my head and pressed me against her breasts. Together we collapsed on the ground, clinging to each other.

"'Stop it! You'll wake everyone up!'

"I breathed through the fabric of her sweatshirt, through the rough Hard Rock Cafe logo. I was aroused and I was afraid she would notice, so I slid away.

"'You're really out of control, Blade,' she said, letting me go. 'Who knows what the hell kind of planet you came from.'

"Then winter came, and it rained constantly. The leaves on the vines gave up and dropped off, done in by all that water. Sometimes I would take solitary walks in the vineyard, and when I was far enough away I would start singing. I would sing the 'Agnus Dei' and the 'Salve Regina' among the dripping vine shoots, my rubber boots sinking into the mud, and I would think of Bern. Now and then I received letters from him, but

I had a hard time answering. I told him about the actress's wedding, about my new duties and how quickly I had learned them, but not much else, because compared to his words mine sounded childish. Bern never tired of writing to me, though, as if he sensed my difficulty. It's ridiculous when you think about it: cell phones already existed and there we were, separated by less than thirty miles, exchanging letters.

"At the bottom of his letters he always asked the same questions. It took me a while to realize that they weren't really addressed to me. 'Do you still believe, Tommaso? Do you believe without having to be forced to? And do you pray, at night? For how long?'

"But all of a sudden God disappeared from his letters. There was no trace of Him anymore. I thought of asking him what had happened, but again I didn't have the nerve. I was worried about him. I knew that there is no deeper solitude in the world than that of those who once believed and then stopped believing. And I had never known anyone capable of believing as absolutely as Bern. Even Cesare, compared to him, seemed to have doubts.

"It was school. It was the people outside. In September, Bern had taken the exam to enter the last year at the scientific high school in Brindisi. From what he told me, he had left the teachers speechless, quoting a passage from the *Metamorphoses* from memory. Cesare had made us study Ovid in depth because, according to him, the poet anticipated the doctrine of reincarnation.

"After Latin, however, came math. Bern had not been able to transcribe on the blackboard the expression that was dictated to him. Sines and cosines? No one had told him about them before. 'This is a scientific high school, Mr. Corianò,' the math teacher had said. 'Didn't anyone explain that to you?'

"In the end, he'd been placed into the third year. Two years' difference was enough to make him stand out among his classmates like a sore thumb. There was a gang of boys who'd grown up on the outskirts of

Brindisi. I knew those people, I'd been raised in the same neighborhoods, but Bern knew nothing about them. He became their target. I could picture him clearly, trying to ward them off with words, as ingenuous as Jesus among the Doctors in the Temple. 'They're full of rage, I feel remorse for them,' he once wrote. He still used expressions like 'I feel remorse for them.' Imagine, with those guys.

"'Don't get involved,' I begged him, but he didn't listen to me.

"I never knew how they tormented him, whether there were only two of them or maybe five or the whole gang. Bern was sure that with patience he would win, that they would finally get tired of him. Instead, he was the one to give up a few months later, because of those boys and because some of the professors didn't give him a chance, there were too many things that he had never studied or that he'd studied differently. 'School isn't for me,' he wrote to me in January, 'much better to learn on your own.' Then he made an effort to talk about things that had nothing to do with it, about the new boy whom Cesare had taken in at the masseria, Yoan, about how taciturn and fearful he was. They had picked olives together on your grandmother's land. 'They're juicier than ever this year' is how he described them. At the end of that letter he lapsed into despair for a moment: 'I miss you very much. I still pray, but most of the time I don't know what I'm praying for.'

"'Talk to Cesare,' I replied, 'confide in him, he'll know how to help you.'

"His answer came by return mail, just one line: 'Cesare kicked you out. I want nothing to do with him anymore.'

"In order not to arouse suspicion, he went out just the same every morning, but instead of taking the bus to school he went on foot to Ostuni, cutting through the fields. He sat in the municipal library all day. He got it into his head to read every book it contained, in alphabetical order. That was just the kind of plan that he would undertake, the same as when he got the idea of living in the trees like the rampant baron, or when he persuaded us to eat the seeds and roots and leaves of every plant that grew on the masseria, or when he dragged me into the strike for the computer.

"He stuck to that plan for three months, during which he wrote rarely, and always about books. He would have made it to the G's, even beyond, but ultimately he became friends with the librarian and allowed him to divert him from that plan and propose a new one.

"'He's introducing me to authors I had never heard of. We were in the dark about so many things, Tommi! And now I'm questioning everything, everything! From the foundations. It's like being born again. We're reading Max Stirner. Every page opens my eyes even more. We lived in darkness, my brother.'

"Thinking to himself, he summed up everything that had gone through his head starting with that book. He called it The Ego. Only afterward would I discover that this wasn't the complete title, but Bern was no longer aware of what I might know, and anyway it wouldn't have mattered much to him. He began signing himself 'The Great Egoist.' In huge letters, in the center of the page, he wrote: 'Our task is to storm the heavens!' He wrote: 'We must devour it, the sacred!' By then he wasn't really speaking to me at all, and realizing that made me feel more alone than ever. In the final letter, before a silence that lasted a long time, he included a passage that was the result of all that study: 'It wasn't I who couldn't pray, Tommaso. I understand that now. I wasn't the one who went wrong. God is a prosaic invention. For only he who is alive is in the right.'"

"HIS COPY of the book is still here," Tommaso said, raising his head. He pointed to a spot on my left. "There, on the shelf."

I got up too quickly, I felt dizzy. Medea lifted her snout at the same instant, on the alert. Seeing me go toward the shelf, she lay back down. The books were all turned on their sides.

"The spine is . . ."

"I found it."

The complete title was *The Ego and His Own*. I handed it to Tommaso and he leafed through some pages.

"Look at all the places he underlined. Practically every line."

He handled it carefully, like a relic. Then he closed it and placed it sideways on a corner of the bedside table.

"My head is exploding," he said.

"Do you want me to get you a pill?"

"I'm afraid I finished them all. Do you have anything?"

"No."

"Too bad."

He massaged his forehead. When he moved his hand away, it left red marks. Then he started talking again, losing himself ever further back in time, as if I weren't even there.

"Going down to the cellar with Corinne turned into a habit. We'd go there at the end of the shift. We'd talk for a long time, passing that makeshift glass back and forth, then we'd climb up on the tanks. At that point everything became very confused. I wanted to go up again and Corinne would pull me down by the heels. 'Enough, Blade! You want to kill yourself?' But I didn't let her deter me, I inhaled again and again. In the end she always said the same thing: 'You are really out of control.' It was like a signal, it meant it was time to part company, because otherwise we'd be forced to go further, to do something that I wasn't sure I could do. I'd go up the stairs from the cellar first, and for a few days we kept away from each other.

"One evening, Nacci sent for me. 'The workers say you play cards,' he said. He joined his hands together on the desk while I hid mine behind my back, fingers crossed.

"'It's not true.'

"'Don't lie to me, Tommaso. I know very well that everyone needs his distractions.'

"He took a deck of cards from the drawer. 'What do you play?'

"'Skat, bridge, canasta. Also scopa, but not as well.'

"'The workers say you play poker.'

"'And poker, right.'

"'I said you mustn't lie to me, Tommaso. And blackjack, do you know it?'

"I hesitated.

"'Do you know it or not?'

"'You mean twenty-one?'

"It was my father who'd taught it to me. All the card games I knew came from my father. Except for skat: that one came from Bern and the mulberry treehouse.

"'Twenty-one, call it whatever you like,' Nacci said.

"'Then I know it.'

"He pushed the deck toward me, new, shiny, springy cards. 'Shuffle them.'

"I did it the classic way, and Nacci studied my hands.

"'Not like that,' he said, 'the American way.'

"I cut the deck in half, put the cards on the desk, and demonstrated.

"'Can you keep one on top?'

"'That's cheating.'

"'Can you do it or not?'

"I showed him I could, but a card slipped out and fell to the floor. 'Sorry,' I mumbled.

"'You're clumsy,' he said, 'and slow. But you can improve. On Friday nights some friends come over. We like to play a little. I'll pay you an extra day's work. You can keep ten percent of what the bank wins. Deal?'

"I began that same week and continued every Friday thereafter. If it was poker they wanted to play and they needed a fourth, Nacci lent me the money to sit at the table. But in general he and his friends preferred blackjack. They spoke very little. On the other hand, they smoked nonstop and drank Jameson from water glasses. At dawn the pilgrimage to the bathroom sped up; they peed without even having the decency to close the door anymore.

"When they finally dragged themselves out of the living room, worn to a frazzle, I put everything back in place: the green cloth folded into the drawer, the chips in the container. I emptied the ashtrays and rinsed the glasses. Before going back to the dormitory I took a walk as far as the vineyard. Only wild animals were awake at that hour.

"What with the evenings at cards and what I saved of my salary, I put some money aside. One day, with a roll of bills in my pocket, I went to the commercial district in Massafra. I found a body shop that had motorcycles displayed on the sidewalk. They looked battered, but it didn't matter. I showed the owner how much I had and asked him what I could buy.

"'Have you got a license, at least?' he asked skeptically.

"I offered him the money again. If he didn't accept, I would look somewhere else.

"'Right, it's none of my business,' he said then. He grabbed the bills and ran them through his fingers, counting. They were small denominations, ten or five thousand liras, as if I had robbed a tobacconist. That was exactly what he thought, I suppose. 'I can give you that one.' He pointed to a scooter. 'It's an Atala Master. Its papers are all in order.'

"I was getting used to my new life. I had my regular work shifts, nights playing cards, evenings with Corinne, and now the Atala as well, to drive around whenever I wanted. I could live like that. I could have lived like that. But then Bern showed up in the courtyard, out of the blue, with his black boots and muddy pants, as if he had waded through a swamp. Seeing him, I tightened my grip on the handle of the basket I was carrying. 'What are you doing here?'

"I set the basket on the ground. I wanted to hug my brother, but I waited for him to make the first move. He stayed where he was.

"'I came to liberate you,' he said, 'get your things, we're leaving.'

"'Leaving? Where are we going?'

"'I'll show you. Now hurry up.'

"Nacci appeared at that moment. I explained to him that Bern was a friend of mine. He glanced at the driveway, and when he didn't see any cars, he asked, 'How did you get here?'

"'On foot.'

"'On foot from where?'

"'From the Taranto station.'

"Nacci burst out laughing, but stopped when he realized that Bern was serious. 'Now I know who you are,' he said. 'You're Cesare and Floriana's nephew. They always said you were a character.'

"He insisted that Bern stay for dinner. It was the only time I ate at his house, though he was deep in conversation with Bern the whole time.

"'Take this boy away,' he finally said to me, getting up from the table, 'he can no longer stand up. And you, give my regards to Floriana and Cesare.'

"As soon as we heard the TV go on in the next room, Bern jumped up. He dumped the remaining bread into a napkin, then the leftovers he had in his plate, as his eyes urged me to do the same. He opened the refrigerator, grabbed a few cans of Coke and a container of yogurt, and hid them under his sweatshirt.

"'What are you doing?'

"'Just these. And these,' he added, taking a carton of eggs.

"'We can't, Bern!'

" 'No one will notice. There's plenty of stuff here.'

"We slinked out of Nacci's house and fled to the workers' quarters. Bern paused in the doorway to check out the room.

" 'My bed is over there,' I said, pointing it out to him, but he no longer seemed interested.

" 'Get a move on,' he said.

" 'I can't go back to the masseria,' I said. 'Cesare was clear.'

" 'We're not going to the masseria.' He took a step and it looked as if he were about to drop to his knees. He clung to the doorframe.

" 'What's wrong?'

" 'Nothing, a spasm in my back. I'll just sit here a moment.'

"He lay down instead, perpendicular to two joined beds. He stared at the ceiling, breathing hard. His T-shirt was raised a few inches and I saw how thin he was.

" 'What happened, Bern?' I asked.

" 'He destroyed all my books.'

" 'Who destroyed them?'

" 'Cesare.' He paused, but I knew he would go on. 'One evening he came into our room and knocked them off the shelf. He ripped them in half, one after the other. They weren't mine, they were from the library. He said, I'm doing it for you, Bern, let the Lord free you!'

"A tear formed in the corner of his left eye. I lay down next to my brother, my head very close to his. He turned his face toward me. When he spoke again I heard his rasping breath: 'We haven't spoken since that night. And I swore to myself that I would never speak to him again.'

"We didn't say anything more. We held hands instead, not talking.

"On the motorbike he clung to my waist, and at one point he rested his head on my shoulder. Reaching out an arm, he opened his hand as if he wanted to stop the flow of air coming at us. The plastic bag in which we'd crammed the food leftovers fluttered in the wind.

"I had never driven for so long. By the time we got close to Speziale, my arms were aching. But Bern said: 'Keep going toward the sea, let's go to the Scalo.'

"'It's spring, there won't be anybody there.'

"'Let's go there.'

"So we made a loop around the hill in Ostuni. The sight of the city surprised me, as if I had forgotten all about it. I loosened the grip on the brake and let the descent alone carry us down to the coast. We rode the Atala to where the brush began, then continued on foot. The trail was barely visible, but Bern moved confidently, headed toward the tower. He lifted the fence post and set out through the nettles. Turning on a pocket flashlight, he traced a flourish on the wall. 'Do you remember how to do it?' He climbed up first. Then he lit up the area below so I could clamber up too. I scraped my knee on a projection. It was all exactly as I remembered it from the summer, but without the reassuring sound of the music outside. In the silence, the interior of the tower was eerie. When we were almost at the bottom, a glow reached us.

"'We're here,' Bern announced.

"I was about to reply that I knew, when I realized he wasn't talking to me. In the room, lit by a battery-powered lantern, were Nicola and a girl. They were sitting on the mattress, her legs crossed, his stretched out.

"'Ciao, Tommaso,' Nicola said, as if finding ourselves down there wasn't at all a surprise.

"'Is he the third one?' the girl asked, though she made no move to get up or offer her hand. Instead she reached for the plastic bag. 'What did you bring?'

"Bern tossed the bag onto the mattress and she started rummaging through it like a savage. 'You didn't get any Snickers?'

"'Whatever was there,' Bern replied vaguely. Then, turning to me: 'Violalibera likes Snickers. If you can, get a few next time.'

"'So he's not staying here?' Nicola asked.

"'No, he likes it where he is. With olive trees pruned so that they look like ornamental sculptures.'

"Violalibera asked: 'Is it true that TV actresses come there?'

"I nodded, but I was still confused.

"'What are they like? Do they have huge boobs?'

"Nicola sniggered.

"'They're normal.'

"'What, you don't like actresses?' Violalibera asked. She was wearing a small headband that lifted her curly hair into a crown. She had a lot of hair. 'Are they much prettier than me?'

"Bern said: 'Violalibera arrived here a month ago. You expected to enter a place like this and not find anyone, or at most a rat, and instead.'

"'Actually, there was a rat,' the girl said.

"Bern ignored her. 'And instead, I come in and get frightened to death. Violalibera was sleeping in pitch darkness and I lit her up with the flashlight. When she woke up and saw me she wasn't even scared, not a bit.'

"Meanwhile, he too had flopped down on the mattress, very close to her. I was the only one still standing.

"Violalibera gulped down a yogurt from the container. She licked the inside obscenely. Mildew: that's what the refuge smelled like. Bern had placed a hand on her leg, up on her thigh. If he'd spread his fingers he would have touched her crotch. She opened the other container of yogurt, ate some, then handed it to Nicola. 'There's no room for one more,' she said.

"Bern may have squeezed her leg harder. 'I already told you he wasn't staying.'

"I felt dizzy, I needed to sit down, but there wasn't enough space on the mattress and I didn't want to sit on the ground.

"'You sleep here?' I asked Bern.

"'When I feel like it,' he replied. 'We can live as we please.'

"Nicola smiled, his teeth gleaming in the lantern light. There was something different about him, a kind of excitement.

"'Are you all pale down there too?' Violalibera asked.

"'Even paler than that,' Nicola answered.

"'So, he's the third one,' she said again.

"Bern unwrapped one of the bundles of leftovers; the grease had soaked through the paper. 'Take some, eat,' he said, and the others went at it with their hands.

"'And do you sleep here?' I asked Nicola.

"'Only if I don't have a class the next morning.'

"'We can live as we please,' Bern said again. Then, from a pile of odds and ends, he produced a tape player.

"'Start it at the beginning!' Violalibera said.

"Bern rewound the tape. The music started, distorted because the cassette was worn and the speaker tiny. Violalibera jumped up, reaching one hand out to Bern and the other to Nicola. Obediently, they got up and started to sway at her sides, practically glued together. Nicola buried his nose in her hair, behind her ear, maybe kissing her, and she shrugged him off, ticklish.

"She brushed my scraped knee with her toe.

"'You, what are you waiting for?'

"Now Bern had a hand on her stomach and waved the other one just over her head. I went over to Violalibera and she pulled me to her. Nicola and Bern made room for us. I inhaled the scent of her hair and the lingering tang of yogurt that came from her mouth. Then the others closed in around us. 'I should . . .' I murmured, but that was all I had the strength to say.

"Bern whispered to me, 'No one gives us orders anymore.'

"Then someone started to undress me, or maybe I did it myself. We undressed one another, as the music grated on the walls. We collapsed on the mattress in a tangle.

"I found my face very close to Violalibera's breasts. Nicola, next to me, was sucking one of them and I felt I had to do the same. Bern slipped between us, every part of us was in contact, he and I, I think I was paralyzed for a few seconds.

"We attached our mouths to Violalibera's nipples in turn, as if drinking from a fountain. Someone, maybe she herself, took my hand and moved it lower, and down there I found that all four of us were naked and excited. I let myself be guided, shifted, then the hand that had steered me was lost and I went on blindly, until I found Bern. I was terrified when I gripped that forbidden part of him, just as I had imagined doing countless times, certain that it would never happen. But he didn't notice, we were so entwined with one another, or maybe he did and let me do it. I knew he wouldn't have let me if the two of us were alone, but in there, in the tower, everything was permissible.

"Before he broke away he smiled at me and I caught my breath. All I could see of Nicola now was his broad, smooth back, bluish in the gleam of the lantern. His face was still buried in Violalibera, who was breathing harder and harder, her arms flung out and her eyes wide open, staring upward. She looked limp, as if she could no longer put up any opposition, as if we three were one huge tangled animal, an animal with many heads and legs and arms, which crawled over and inside her, making her moan.

"For a moment I looked toward where she was staring, but there was only the gray ceiling. I pictured what still existed beyond those walls, the nettles, the rocks polished by the sea foam, darkness everywhere. But down there none of that mattered anymore. We were protected and alone, unreachable. And I wished it would never end."

FOR A MOMENT we returned to the present: Tommaso and I as adults, Christmas Eve, his daughter asleep on the other side of the wall. He threw his head back and I instinctively looked up at the same point of the ceiling, but there was nothing there, except for the glowing halo cast by the lamp. I knew that was not what he was seeing.

I shifted in the chair. I felt a kind of nausea, nausea and disbelief, and something less confessable: envy because I had not been with them in the tower? For a moment I thought of telling Tommaso enough, that he should keep the rest to himself. What was the point of knowing all that now? But he slowly straightened his head and I let him continue.

"It's wrong, I told myself, what we're doing is wrong. It's immoral. But I went back there whenever I could. It didn't happen often, though, four, maybe five times in all. Yes, five at the most," he repeated, "maybe six. When I finished my work at the Relais, I got on the Atala, cut across Martina Franca, and swung down to the coast. I wanted to get to the tower as fast as I could.

"'There's some girl involved,' Nacci said when I asked his permission to leave yet again. I didn't answer. It wasn't a lie, after all. 'A wonderful age, yours,' he added, 'once in a lifetime,' then he took a bill out of his pocket: 'Take her to a restaurant.'

"I used the money to buy some pasta and pancetta, Snickers, and two bottles of Primitivo. We cooked on the camp stove, near the stairs, so the smoke would at least partially clear out. Outside the tower the days were lengthening, making me edgy. By then I preferred the perennial night of the tower, the cold gleam of the lantern.

"That evening we smoked a hookah that Nicola had bought at a flea market in Bari, with apple tobacco; we blew it at one another. Then we made hand shadows on the wall: Nicola was a dog in profile, Bern a mouse, I a bat, and Violalibera a peacock. Our animal silhouettes touched, teased one another on the wall. But we were worse than animals, we four.

"One day, during a reception, Corinne grabbed me by the sleeve. I almost tipped over the antipasto tray.

"'So you don't feel like getting high anymore?' she asked.

"'No. I mean, sure. Why?'

"'You haven't come down to the cellar anymore.'

"'I'm just a little tired.'

"'Where do you go all the time?'

"'Nowhere.'

"The guests had already taken off their shoes and were wandering around the lawn among the sago palms.

"'They say you have a girl in Pezze di Greco,' she persisted.

"'And you believe it?'

"'Why shouldn't I believe it?'

"She couldn't contain her bitterness.

"'It's not like that,' I said softly.

"'After all, there's no reason why I should care, right?' she hissed. Then she looked me straight in the eye. 'Right, Blade?'

"She crushed out her cigarette in the crack between two stones in the wall.

"'Suit yourself,' she said, bumping my shoulder as she passed me.

"The tray really did overturn this time. The little glass cups with the shrimp in cocktail sauce scattered on the floor. I placed those that weren't too damaged back on the tray and served them to the guests.

"Later I wrapped up the party's leftovers: meatballs, small squares of eggplant parmigiana, and batter-fried vegetables that would get mushy in the paper napkin, but that we would eat anyway, soggy and cold.

"It was June, at the Scalo the season was about to begin. The tables and benches were already stacked on the flat stretch of rocks, the pink trailer-bar silhouetted against the sea.

"By then I could climb the tower and go down the steps without turning on the flashlight, feeling the crumbly walls.

"'Fried food,' I said, yanking the backpack off my shoulder. Then I repeated it, because nobody had answered me.

"I noticed Nicola first. Sitting on the mattress, he was holding his head in his hands. He didn't turn toward me. A moth was flapping against the lantern. Who knows how it had gotten down there.

"Bern was lying on the floor, on his back, hands crossed over his chest. I took the bag of leftovers out of the backpack and waved it in front of him.

"'Leave him alone,' Nicola said, 'his back hurts.'

"Bern didn't move, his eyes were closed. If I hadn't been sure he'd stopped praying a long time ago, I would have sworn that's what he was doing. Though maybe he really was, just that one time, or so I tell myself today.

"'Where's Violalibera?' I asked.

"No one answered me. I was relieved not to find her there. For once it would be just the three of us, like before.

"I looked at Bern, who was now pressing his chest with his fingers.

"'It must be the dampness,' I said, 'it got into your bones.'

"'How much money do you have?' Nicola asked me.

"I rummaged in my shorts pocket, opened my wallet under the lantern. 'Fifteen. Why?'

"'How much more do you have set aside?'

"'I always bring you stuff to eat.'

"No one else contributed. Nicola spent his parents' monthly allowance on living expenses in Bari and on gas; Bern and Violalibera had nothing.

"'Doesn't that guy pay you to play cards?'

"'I don't play cards. I'm the croupier.'

"I realized only then that Nicola's eyes were wet. When I told him how much I had, not the real figure, about half, he put his head in his hands again.

"'What do you need it for?'

"No one answered. The moth had alighted on the brightest spot and seemed to pulsate. Finally, Bern spoke to the ceiling, his voice strained: 'Go on, Nicola, tell him.'

"'Why don't you tell him?'

"'So, what do you need it for?'

"'Looks like we messed up,' Nicola said. His laughter was explosive in the small space, completely unexpected. 'Messed up but good, yeah.'

"He abruptly stopped laughing and started trembling. The moth resumed flying nervously in circles, grazed my face.

"'She's pregnant,' Bern announced from the floor.

"When Nicola had calmed down, his eyes bored into mine. 'Could it have been you? Maybe it will have white eyelashes.'

"Slowly Bern heaved himself up. He crossed his legs and tried to move his shoulders. The spasms that seized him, he said, started at his temples and came together like two branches of a river and coursed down his spine, to his groin. They lasted as long as a week. But you already know this.

"'Let's go out,' Bern said.

"I helped him stand up, then climb the stairs. Where the steps were missing, he let himself slide. We walked through the nettles and sat down on the trailer hitch. Bern spoke the words slowly.

"'There's this doctor,' he said, 'in Brindisi. He does everything without anyone finding out. But they say he wants a million lire.'

"I asked again where Violalibera was. Nicola started crying. Bern looked at him coldly.

"'For now we have two hundred thousand,' he went on. Only half of his mouth moved. 'Floriana will give Nicola the same amount next week. Adding yours, it comes to almost five hundred thousand.'

"Nicola was in a panic. 'Do you remember what Cesare says? Huh? Do you remember?'

"I was worried someone would hear us if he went on talking so loudly, but there was nobody around for miles, only the geckos hunkering under the scrubwood, only the crabs wedged in the cracks between the rocks.

"Bern gripped his arm, but Nicola shook him off. 'What happens to babies who are killed before they're born, do you remember that?'

"'You're acting irrationally. There is no reincarnation, no punishment, no divine being. We've already talked about it. If you had read *The Ego* . . .'

"'Shut up! It's because of that fucking book that we ended up like this!'

"'The fish,' I murmured.

"Some tribes, Cesare used to tell us, threw dead newborns into rivers because they didn't yet have a soul, and without that they would not be reincarnated. They gave them to the fish to eat so that a soul would find them there.

"'We'll be damned,' Nicola whimpered.

"Anyone who doesn't welcome a visitor will be reincarnated as a tortoise, Cesare used to say. Anyone who kills a large animal will go insane. Anyone who eats meat will be red, a ladybird or a fox. Those who steal will slither. Whoever kills a human being will be reborn as the most abominable of creatures, that was what Cesare used to say. Then he'd say: pray to the Lord God that He may have mercy on you, pray incessantly for His forgiveness.

"'I have two hundred thousand. I lied before. At the Relais I have two hundred thousand liras,' I said.

"'Then that makes six hundred. We need four hundred more.'

"'Maybe two hundred and forty, I don't know. I have to count it.'

"Nicola jumped to his feet. 'Didn't you hear me? Don't you remember anything anymore? God will abhor us for it! God already abhors us!'

"Again Bern answered him calmly: 'There's always a second solution if this one doesn't satisfy you.'

"Nicola looked around for a moment, bewildered. He took a few steps away from us, then stopped. All that emptiness around us reflected his dismay.

"'You see?' Bern said. 'Only we exist. The great egoists. There is no god who can abhor us.'

"His serenity scared me almost more than Nicola's despair. With some effort he added: 'Everything that Cesare told us is a lie. Human life is only . . . ,' but Nicola leaped at him and started shaking him.

"'Cesare is my father! My father, understand, you bastard? You're the only liar, Bern! Look at what you got us into!'

"I grabbed him by the neck, until he had to let go of Bern to pull me off. When I stopped squeezing, he began coughing.

"'We'll find the money we need,' Bern promised.

"At that point a weariness suddenly came over us. I looked at the expanse of rocks and only then did I see a figure standing on the shore, a shadow barely darker than the rest, very far away. Violalibera.

"Closer to us were the sandy patches where we had danced the summer before. But who remembered that anymore?

"The night is made for sleeping: that's another thing Cesare used to say, before turning off the light in our room, before giving us the last whispered blessing in the shadowy dimness. The night is made for sleeping. But not for us, we didn't want to sleep. So we listened to his footsteps move away along the corridor, then we turned on a flashlight and climbed onto Bern's bed. On that raft we continued our games until it was late, the games children played, games that were innocent yet each night a little more daring, each night a little more risky.

"All of a sudden I saw the figure on the rocks dive off. The sound of the plunge could barely be heard. Unable to move, I said: 'She jumped into the water.'

"Bern and Nicola quickly turned and ran toward the rocks, calling out Violalibera's name. Then I too followed them. We stood on the edge of the rocks, all three of us shouting now. The waves were crashing, froth flying. Luckily, there was a little moonlight. Nicola pointed to a spot in the water: 'Out there! She's over there!'

"He didn't have the guts to jump in, but Bern did, diving in feetfirst without looking at what was below.

"'Shit!' Nicola cried.

"I dove in too. The water was so cold it knocked the breath out of me. I hit something on the bottom, resurfaced, and swam toward Bern, who had meanwhile grabbed Violalibera and was holding her head above water.

"Then Nicola also joined us. We clung to her, until she said, 'Okay, let go of me! Let go of me!'

"We swam to shore and helped one another climb up the rocks. The current pushed me back twice before I managed to drag myself up.

"I was trembling from the cold. Violalibera said we had to take off our clothes or we'd catch pneumonia; we obeyed her, and took everything off. Then she told us to huddle close to her to warm her up and again we obeyed. She started laughing then. 'I gave you a scare, didn't I?' she said, and meanwhile, she wiped the drops off our skin with her hands and lips and hair. I found myself kneeling on the sharp rocks, then lying on my back. Fear had excited us. I looked up one last time. Even with the moon you could see countless stars.

"The next day Nicola was waiting for me in front of the cathedral in Brindisi. 'Leave the motorbike here,' he said, 'we'll walk.'

"'They'll steal it if I leave it here,' I said, but he had already started walking. I shut off the motor and trudged behind him, pushing the Atala. We headed for the waterfront. It was strange to be there alone, just us two, in broad daylight. All of a sudden Nicola said: 'I've thought about it. Bern has spent more time in the tower than we have. Much more time.'

"'And what does that mean?'

"'Nothing. Only that he's been with Violalibera more. It's a fact. How do we know what they do when we're not there?'

"'We were there too, Nicola.'

"'I'm sure it couldn't have been me.'

"'You can't say that.'

"He gave me a nasty look. 'Sure, you always defend him. You can't even see what he's become.'

"'What has he become?'

"'A fanatic, that's what. And only to provoke Cesare.'

"'Yeah, but Cesare . . .'

"Nicola stopped short, we almost collided. 'Cesare what, huh? You two are always ready to attack him. He took you in and took care of you. Without Cesare you would be . . .' But he didn't finish the sentence.

"'He destroyed all his books.'

"'All his books? Is that what he told you? Two books. Only two.'

"'Two,' I repeated softly to myself. I tried to remember the details of the conversation with Bern in the bunkhouse at the Relais. But did two or a hundred make a difference?

"In the meantime we'd arrived at the address, on a street in the old city. One of the balconies was brimming over with the shoots of a succulent, which had wound itself around the railing like tentacles. Nicola checked the house number on the slip of paper he had folded in his pocket.

"'This is it,' he said. 'You buzz.'

"'Why me?'

"'Buzz, dammit!'

"An old woman came to open the door; without saying a word, she stepped back to let us in. She pointed to the sofa with a weary gesture, then sat down in a nearby chair and went back to watching television, an afternoon variety show. I had made up an excuse for Nacci, again regarding the nonexistent girlfriend. Seeing the actresses on the screen, for the first time I thought about the Relais with some regret. About Corinne.

"'Come in,' a man's voice behind us said.

"He had a thick, well-groomed beard, and wore a pair of rimless spectacles. He led us into the kitchen. 'Where is the girl?'

"'She didn't come today.'

"'So I'm supposed to examine one of you?'

"'We didn't think . . . ,' Nicola said, but then, embarrassed, he fell silent.

"'How many weeks is she?'

"'A few. We think,' I replied stupidly.

"'Which of you is the father?'

"This time we both remained silent. The doctor turned to the sink, filled a glass from the tap, and drank it in one go. Then he put the glass back on the drainboard without rinsing it. He didn't offer us anything. 'I see,' he muttered. 'She's a minor, correct?'

"'She's sixteen.'

"'You must bring her to me as soon as possible, is that clear?' He seemed both weary and disgusted as he said it. The drone of the television could be heard from the other room. 'The operation costs a million and a half,' he added.

"'We were told a million,' Nicola said, suddenly nervous.

"The doctor smiled grimly. 'You don't know who the father is and you don't know what week she's in either. But you knew the price, right? Well, the price is different. If it can't be done, I will give you back a million three hundred thousand. I will only deduct the cost of the examination.'

"'What do you mean, if it can't be done?' Nicola repeated his words.

"'Doctor,' I interrupted.

"'Tell me.'

"'How is it done?'

"He stared at me for a few seconds. Then he turned, opened a drawer, and took out a knife. He held it up so I could see it plainly, laid the serrated blade against the tabletop and scraped it, as if scraping something off, a patina. 'Is it clearer to you now?'

"Nicola had turned pale.

"'It was you,' the doctor said, 'not me.'

"Back at the Scalo, we forgot about eating. The music from outside reached us muffled. Violalibera took a piece of paper and set fire to it with a match, then let it burn in her fingers. The flame that was quickly consumed gave the most light I had ever seen down there, and for a few seconds it fully revealed our terrified faces.

"We counted the money once again: nine hundred thousand. At that point, I had exhausted every reserve.

"'We'll never make it,' Nicola said. I was afraid he'd have another attack of nerves.

"'You could borrow some,' Bern told him.

"'Oh, really? From whom?'

"'From your university buddies. They must have money, those guys.'

"'Why don't you do something about it instead? You sit here always giving orders without doing a fucking thing.'

"Bern smiled. 'I see that your law studies are enriching your vocabulary.'

"Violalibera started snickering. That night she was wearing a tank top that exposed her navel. She stretched her bare foot toward the center of the room, toward Nicola, rubbed it against his thigh and then against his crotch.

"Nicola grabbed her foot as if he wanted to mangle it, then shoved it away. 'You're crazy.'

"That's when Bern turned to me. His back had neither improved nor worsened, but he no longer complained. Now his attention was constantly on Violalibera; he took care to make sure she had enough to drink, that she was comfortable. He hadn't returned to the masseria anymore, so as not to leave her alone. I didn't know what Cesare and Floriana thought of it, whether they were sick with worry; he didn't talk about it. He'd made the crumbling interior of the tower his new home. And instead of lying on the mattress next to Violalibera, he slept on the floor to leave her more room, his aching back on the hard ground.

"'You have to get the remaining money we need,' he said.

"'How?'

"'They must keep money at the Relais.'

"'You want me to steal it?'

"He sat opposite me, thin, pale. 'Take it from the cash register the first night when there are a lot of guests. Not too much, leave enough not to raise any suspicions. Be smart about it. If necessary, you'll have to do it more than once.'

"'Bern,' I murmured, 'no. Please.'

"He slid his backside toward the mattress. Sitting beside me, he pressed my head against his shoulder and stroked the spot between my ear and my neck.

"'Poor Tommaso,' he said. 'We are all very grateful for what you do.'

"'Bern . . .'

"He tapped the back of my neck softly. 'You know that, don't you?'

"Under the holly oak, at a time that seemed to have nothing to do with the present, Cesare had taught a lesson about the Commandments. '"Thou shalt have no other gods before Me": the Lord had dictated this law to Moses first, but why?' he had asked. 'Why that law above others that to us seemed more important, above "Thou shalt not kill," for example?' Cesare had looked at us in turn. We were silent. So he had answered for us, as he always did: 'Because when the Lord is replaced in our hearts, what follows is nothing but an endless plunge, falling headlong without stopping and violating every other law. When the Lord is replaced in our hearts, we always, inevitably, end up killing.'

"During that period I went to see my father in prison. There was no one else in the visiting room besides us and the guard. He sat across the empty, polished table, a table identical to all the others in the room.

Even sitting still we were sweating. Our hands never touched, it had been like that even before prison. At times I thought my father wanted to, that he would gladly reach out to touch me, but he forbade himself. Yet I would have let him do it, maybe not earlier, but by then I would have let him take my hands and hold them in his.

"'Have you learned to balance and carry the plates?' he asked me.

"'Three at a time. Even four, if someone helps me load them.'

"'Four. I would drop them all.'

"He always wore his shirt when I went to see him, the same one, checked, the first two buttons open. A thin silver cross hung around his neck.

"'You're sad,' he said.

"'I'm okay.'

"'Is it because of a girl? One you met in Massafra?'

"I bowed my head. His fists opened slightly, the blood flowed into his white fingers, but then he tightened them again.

"'Maybe I got her pregnant, Dad. Maybe it was me or maybe it wasn't, but I was there. I wanted the others more than her, but I was down there.'

"As if sensing something, he said, 'Don't worry, Tommaso. You won't turn out like me.'

"Then the guard approached the table. He didn't need to tell us that time was up, or point to the clock on the wall. All three of us knew the drill. I stood up first.

"The following day the garden at the Relais was decked out in an exaggerated display of pink and white. I helped the gardeners trim the box hedges, then I did a final inspection of the tables: silver plates and cutlery, tablecloths skimming the floor, and in the center of each table an arrangement of fresh flowers. I checked that each place setting had a napkin folded into a swan. Some of Floriana's little tasks had unexpectedly proved useful; swanlike napkins inspired Nacci's satisfaction. Toward four o'clock in the afternoon the party began to get wilder. The children ran around the courtyard, the music's volume was turned up, and the guests split up between the dance canopy and the bar. Hard liquors and spirits were not included in the price because they were what Nacci made a greater profit on. Corinne and I took turns serving them. We still weren't speaking to each other after our argument. At a chaotic moment I opened the cash register, grabbed a handful of bills, and stuffed them into my pants pocket.

"The party girl, an eight-year-old who had made her First Communion that morning, began unwrapping her presents. Everyone crowded around her and I took advantage of it to stick my hands in the cash drawer a second time. When I looked around I saw Corinne watching me from behind the window. She didn't shake her head or anything, but she kept her eyes on me long enough to let me know that yes, she knew what I had done. Then she moved off into the courtyard.

"When I pulled out the money in the bunkhouse, the bills were damp with sweat. I didn't count them until I was safe inside the tower with my brothers, that same night. Grabbing the bills blindly, I had stolen less than I thought. But Nicola had decided to ask some friends for a loan. So we now had a million two hundred thousand lire.

"The lantern battery was very low and the light flickered.

"Bern asked: 'When's the next party?'

"'In a week. I think.'

"'Couldn't you have taken more?' Nicola burst out.

"'They would have noticed.'

"'If we wait longer, the doctor won't do it anymore. He told us so.'

"Violalibera looked terrible. I think she sometimes vomited, even though she barely ate anything I brought. I didn't know how many days it had been since she'd stopped washing.

"'Come here,' Bern said, 'all of you. Closer.'

"I went to him, obedient as always. He sat there stiffly, his rigid back supported by the wall. Violalibera clung to his body on the opposite side, then she ordered Nicola: 'You too.'

"'No,' he snapped. 'Don't any of you realize?'

"'Come here,' Violalibera insisted.

"Nicola came over and, as if suddenly surrendering, slumped down, his head on her legs.

"'We've been apart from one another too long,' Bern said. It was as if he held us all firmly in a single embrace.

"That's when I said, 'I'll go to Cesare.'

"'And tell him what?' Bern asked.

"'I'll go to Cesare,' I said again.

"They accepted that promise without a word. Through the bodies, Violalibera reached for my hand. Now everyone was touching everyone else. Wasn't that our game? To bond together with every muscle and nerve? And then explore every inch of her, inside and outside of her? I felt the rhythmic pressure of the blood in her fragile pulse. I wondered if it was the same beating of the thing she had inside her.

"'Amen, I say to you, whatever you did for one of these least brothers of mine . . .'

"But there was no god, so there would be no judgment. The lantern light wavered, the battery nearly drained.

"When I woke up, we were still joined like that. The lantern had gone out. Violalibera's breath brushed my forearm, but the rasping sound was Nicola's snoring. I unstuck my cheek from Bern's thigh, damp with my sweat or his. I carefully freed myself from the tangle of arms and legs and dragged myself up the steps on all fours. When I reemerged, I was struck by the same astonishment as always: the world still existed outside the tower.

"I had driven so many hours in the last few days, back and forth from one coast to the other, that the stripes left on my hands by the scorching rubber of the handlebar might never fade. But the country air on a

Sunday morning was fresh and invigorating. I arrived at the masseria before it was even eight o'clock.

"I had left that place only ten months ago and already I felt like a stranger. The jumbled stacks of wood, the untrimmed foliage of the trees, the cucumbers invading all the other plants in the vegetable garden. By then I was used to the cultivated grounds at the Relais. I was hoping to find only Cesare up, but he was having breakfast under the pergola with Floriana and the other boy.

"'Tommaso, good God! What a surprise! And at this time of day. Come, sit down with us. Have some breakfast. Yoan, bring a chair over to the table, would you? Where on earth did you come from, dear boy?'

"He held me in his arms. There it was again, his body, that warmth different from all the others, the reassuring scent of his aftershave. I sat down. Floriana touched my hand and pushed a plate of sliced bread over to me.

"'Put some butter on it,' Cesare urged. 'We buy it at a farm just past the Apruzzis' property. Just think, not even a mile from here and we didn't know it existed. Yoan passed it by chance. So many things we don't see, when they're right under our noses! They have excellent cattle, nice fat white cows.'

"I sank the knife into the bar of butter, which had softened from the rising heat, and spread some on the bread. I was starving and hadn't even realized it.

"'Put some more on. And some sugar on top. Butter and sugar certainly can't hurt at your age. I'm the one who should be careful, but what can I do? I've always been a glutton.' He watched me bite into the bread and chew it. Then he smiled. 'Of course, who knows what delicacies they've accustomed you to at the Relais. Do you have any news to tell me about Nacci? It must have been last summer that I last talked with him.'

"'There's always a lot of work,' I said, 'with weddings and all the rest.'

"'Nowadays that's what people do, have lavish celebrations. Floriana and I arranged everything ourselves, right here. It certainly wasn't a time when the groom went to have a manicure before the ceremony, if you get what I mean,' he said, winking at me.

"'I have to talk to you,' I said, my voice sounding harsher than I intended.

"'I'm here, Tommaso. I'm listening to you. We still have half an hour before we go to mass.' I watched Floriana, her lips drawn in tightly.

"'I have to talk to you alone.'

"Cesare stood up. 'Of course. So let's go to our place, shall we?'

"We walked toward the holly oak, I a few steps behind. I had hoped he wouldn't take me there, but I focused on what Bern said: none of the things Cesare had told us was true, his words were nothing but illusory games and conditioning. In the world, only we existed. I sat on the faded bench as if it were just another piece of wood.

"'Would you like us to pray together first?'

"My head gave an uncontrolled nod. He began reciting Psalm 139 from memory, with his eyes half closed and the enfolding voice of one time: "'Lord, you have probed me, you know me, you know when I sit and when I stand, you understand my thoughts from afar, you observe my actions and my rest, all my ways are known to you."'

"The words of the psalm triggered an unexpected exaltation in me. I wasn't prepared for it and I struggled to overcome it. For years I'd been ashamed to be the only one at the masseria impervious to the word of God, doubting my ability to feel it as deeply as my brothers. Often, sitting in the shade of the holly oak, I had expressed that uncertainty to Cesare. He had always accepted it with the same response: 'There is no right way to pray, Tommaso, your desire is already your prayer.'

"'What is it that's worrying you so much that it brings you here at this hour of the morning?' he finally asked.

"I took a breath, then I said, 'I need some money.'

"Cesare straightened his shoulders and arched his eyebrows. 'I wasn't expecting that. I admit it, I really wasn't. I thought Nacci was paying you a salary. Isn't it enough? I can talk to him if you like.'

" 'I need six hundred thousand lire,' I said. I don't know why I picked that figure when half would have been enough. But the memory of my last conversation with Cesare, in that very spot, the day he'd sent me away, had suddenly come back to me. Puffing out his cheeks, he held his breath for a moment. 'I really wasn't expecting this,' he repeated. 'Have you gotten into some trouble maybe?'

" 'That's my business.'

"In all those years I had never dared speak to him with that tone; I had never even dared to imagine it. But Cesare did not lose his composure.

" 'You're unpredictable, you boys,' he said, 'a mystery to me. Does our Bern have anything to do with it? He hasn't been around for days. The older he gets, the less I understand him.'

"Had I met his eyes at that point, he would have read the whole truth in mine. So I kept them stubbornly on a rock and a tuft of grass as I pronounced my threat clearly: 'If you won't give it to me, I will tell Floriana everything.' There was a moment of silence, broken only by the call of a bird hidden in the foliage above us.

" 'What will you tell Floriana, Tommaso?' Cesare asked in a low voice.

" 'You know what I'll tell her.'

" 'No. I don't know.'

"I took a deep breath. 'About when you spied on Bern and Teresa in the reeds.'

"Don't look at him, I kept telling myself. Keep your eyes on that rock and the tuft of grass.

" 'I feel so sorry for you, Tommaso.'

" 'Six hundred thousand. I'll come and get it Thursday evening.'

"I was determined to get up right after I said it, but my leg muscles didn't obey me. I sat there, just as I used to do, waiting for his absolution.

"'So you're blackmailing me. This is what you have become.'

"'Thursday evening,' I repeated, and finally I stood up.

"I walked toward the motorbike without turning around. I struggled with the kickstand, flipping it clumsily, and made a half turn onto the dirt track. Only then did I look back at Cesare in the rearview mirror. He was still there, under the holly oak, wide-eyed with amazement. To me he seemed like merely a defeated man, just as Bern said. And yet, all the way back, the faster I went to get away from there, the harder the wind slapped my shame back in my face.

"When I reached the Relais it was raining; it seemed dark as night in the middle of the day. I went into the bunkhouse and immediately noticed Corinne's glass, the plastic bottle bottom, in the center of my bed. I picked it up, not understanding at first. The key to the cellar glinted inside it.

"I went running out again. I raced through the reception hall, not caring about scuff marks on the marble floor. In the dressing room I opened Corinne's locker: empty. Her Reebok duffel bag was gone, her uniform gone, her personal supply of candy gone. I strode into Nacci's office without asking permission. He looked up at me quizzically. 'Apparently someone went out without an umbrella,' he said and chuckled.

"'Where's Corinne?'

"Nacci waved a hand dismissively. 'Gone.'

"'What do you mean gone?'

"'She's a junkie, I thought I told you that. There's no hope for those like her. They never change.'

"The drenched T-shirt stuck to my back made me shiver.

"Nacci sighed. 'She thought she'd take some of the bar earnings. At this point I wonder how many times she did it. But yesterday the shortfall was so considerable that it left no room for doubt.'

"'Did she tell you that?'

"Nacci looked at me again with that puzzled expression: 'Have you ever heard a junkie plead guilty? Though when I asked her she didn't deny it. I told her she could return the money or leave immediately. She chose to leave, obviously.'

"'Corinne doesn't use anymore,' I said faintly.

"But Nacci had turned back to the documents he was looking over. 'What she does or doesn't do is no longer my concern as of . . .'—he glanced at his wristwatch—'two hours ago. Hiring her was a personal favor to her father. A little like in your case,' he added, shoulders shaking as he chuckled to himself, as though he found the coincidence amusing. 'Go dry off, now. We can't transplant the viburnums with the soil all wet like this anyway. Or, better yet, no. Since you're already soaked, the lawn should be sprayed with mosquito repellent. When there's water, the bastards lay eggs.'

"The storm passed on but continued to rumble in the distance. The first rays of sun that pierced through the clouds were scorching. The strap of the jerry can cut into my shoulder and the liquid sloshed from side to side, making me lose my balance. I sprayed every shrub, every flower, every blade of grass with the repellent. I didn't even think about the microscopic massacre I was enacting. Is this what you have become? In the evening, in bed, I rubbed the rim of Corinne's parting gift on my lips. At the end of that watery, messed-up day, I found myself thinking of her with a new longing.

"In the week that followed, when I wasn't working, I lay on the bed looking at the red tips of the apricot branches outside the window. I wondered if Cesare was spending those hours in prayer, invoking guidance, and whether I would really be able to tell Floriana what I had threatened to say. What words would I use? If the plan fails, I told myself, I'll steal from Nacci again, and I'll end up in prison, like my father. I gladly let those heroic fantasies overtake me, then I succumbed to nausea. But on Thursday I made the trip to the masseria with my heart strangely light. I left the motorbike at the iron bar across the dirt track and continued on foot. The pears on the fruit tree had already changed color. It was sunset time, the hour that for years had made me think I could never live anywhere else.

"When I knocked, Cesare's voice invited me in. Once again I hoped to find him alone, and once again Floriana was sitting at the table with him. He told me to sit down and offered me some wine, which I refused. Floriana didn't even say hello.

"'So you came back for the money,' Cesare said. Then, when I didn't answer, he added, 'Am I right?'

"'Why don't we go outside?' I said. But he ignored my suggestion. 'I can't give you the money, Tommaso. I'm sorry. I spoke to Floriana, I told her everything. And I have to thank you, you know? Without your intercession I wouldn't have had the strength; I would have gone on carrying the weight far too long. Shame provokes the worst in each of us.'

"'You're just a greedy little bastard!' Floriana burst out.

"Cesare touched her arm to calm her. He closed his eyes and murmured something to redeem those words instantly. Then he said: 'Now you too have the same opportunity, Tommaso. Tell us what happened. Maybe we can help you.'

"But I couldn't stay there any longer. I ran out of the room, then through the yard and down the dirt track. I climbed on the Atala and drove off.

"At the Scalo I didn't take any precautions to get to the tower. In there I found Bern and Violalibera asleep. They spent almost all their time like that, because staying down there made them sluggish. The lantern was on, maybe Nicola had brought new batteries. I yanked at Bern's filthy T-shirt and he struggled to open his eyes.

"'Tommi,' he said.

"'He won't give us the money.'

"His lips were parched, his breath rank. I felt his forehead. 'You have a fever, Bern.'

"'It's nothing. Help me get up. Today my back doesn't want to obey me.'

"Violalibera was still sleeping, lying on her side on top of the mattress.

"'Do you have some change for a couple of beers?' Bern asked. 'I could go for one. I'd like to go outside awhile.'

"But we stayed in the tower for quite some time before deciding, speaking softly, or maybe we were silent.

"It definitely wasn't a short time, because as I was finally helping him stand up, his body burning with fever, Cesare appeared in the room. As if the shadows had produced him.

"'Bern,' he said.

"Bern tried to break away from me and almost collapsed on the ground. I held him up. 'Why did you bring him here?' he asked me, his voice full of sadness.

"'I didn't bring him.'

"'Let me help you, Bern,' Cesare said, and took a step toward us. He put his arms around my brother's waist and Bern submitted to that embrace with such abandon that I thought he might have fainted.

"'Forgive me,' Cesare whispered in his ear.

"Yoan must have been lurking somewhere and must have followed me when I fled. Once at the Scalo, he'd called Cesare. And now there he was. Bern was sobbing against his chest.

"There was no need to explain the presence of Violalibera, who had meanwhile awakened. Cesare did not ask questions; all he said to us was: 'Come with me. I'll take care of you.' He bent over Violalibera and stroked her distraught face: 'You too. Come.'

"And so we followed him docilely, up the first set of steps and down the second. Through the nettles, he supported Bern on one side and Violalibera on the other. Before leaving the tower I stuck the money we'd scraped together into my pocket. We passed among the young people at the Scalo, some of whom greeted us.

"We piled into the Ford, and Cesare drove to the masseria without saying a word. Or actually, he did say one thing, just a few words, addressed to Violalibera: 'You'll like it where we're going.'

"That's when I thought: he already knows.

"As if the plan were even more extensive than I'd thought, Nicola was waiting for us at the masseria along with Floriana. Only now does it occur to me that he himself could have been the one who told Cesare where to find us. Oddly enough, I had never considered that. If there was one person from whom Cesare could extract the truth, it was Nicola. Whatever the case, from the pergola Nicola gave me a pointed look, a look that I remember clearly.

"Floriana phoned the doctor in Speziale to come and examine Violalibera, even at that hour, yes, right away. Bern, Nicola, and I left her in their care and slipped away from the house. We walked to the center of the olive grove and there the full brunt of Nicola's panic exploded at me.

"'What did you tell him? What the hell were you thinking?'

"'He didn't tell him anything,' Bern answered for me. 'As for Violalibera, he must have figured it out on his own.'

"'You have to keep me out of it, guys. Leave me out of it, please. I'll give you whatever you want,' Nicola begged us. His face was distorted by dread.

"Bern ordered him to shut up and his tone was so firm that Nicola fell silent.

"Then Bern added, 'We have to decide which of us is the father. When the doctor comes he'll want to know. Cesare and Floriana will also want to know who it is.'

"'Not me,' Nicola whimpered.

"Bern searched around for something. 'Here's what we'll do,' he said. 'We'll each pick up a stone, one of these. We'll throw them toward those trees. Whoever throws it the shortest distance will declare himself the father.'

"'You're completely out of your mind!' Nicola shrieked.

"'If you have a better suggestion, I'm listening. No? I didn't think so. Let's find the stones then, and they must be roughly equal in size. Like this one.'

"I found mine and rubbed a little soil off its surface with my thumb. 'What if it isn't true?' I asked. 'What if the one who loses isn't really the father?'

"'The truth is dead,' Bern replied impassively, 'it's a letter, a word, a material that I can use up.'

"'What if Violalibera doesn't agree?'

"'She already agrees. But we have to swear an oath.'

"'What oath?'

"'Once the contest is decided, none of us will ever again speak of this moment, nor of the tower. We won't talk about it with others and we won't talk about it among ourselves. Ever again.'

"'All right,' I said.

"'You both have to say it: until death.'

"'Until death,' Nicola swore.

"'Until death,' I too swore.

"'Nicola, you go first.'

"Nicola expelled the air from his lungs, filled them again, arched his back, and hurled the stone, very high and very far; it landed beyond the third or fourth row of trees. It bounced dully once, then became invisible.

"'Now you, Tommaso. No, take this one.' He put a stone in my hand that was different than mine, smoother.

"'It doesn't count if you help him,' Nicola protested, but he quickly fell silent. After all, he knew I wouldn't be able to match his throw. When I saw my stone plummet a mere twenty yards away I wondered if the test might be a trap. I had always been the most inadequate in that type of contest. But I was also the one who had never rebelled against Bern's decisions. For the first time since the afternoon I'd met him under the mulberry's branches, I hoped against him.

"I'm not sure if he did it on purpose. Whether it was because of his back, the fever. Or whether it was purely a blunder. I don't know. And our oath would prevent me from asking him for the rest of our days.

"Bern raised his hand over his head, and a sharp pang seemed to freeze him in that position; then his hand let go of the stone. It dropped just past the closest olive tree. We were silent, all three of us. We gaped at the spot, just as long ago we had gaped at the wooden cross that had appeared on the hare's grave.

"Then Bern said: 'I guess I'm it.'

"When we returned to the masseria, he went over to Violalibera, who was staring at the empty plate in front of her. He put a hand on her shoulder, and though she did not react to the contact, that gesture was enough to make clear to Floriana and Cesare how things were, which of us bore the blame. Cesare placed a chair beside that of Violalibera, for Bern. Then he did what no one could have imagined. He left the house and returned a few moments later carrying a basin, the same one we once used to keep the melons cool. He filled it from the sink, then set it on the floor in front of Bern and Violalibera. He took off their sneakers and socks and placed their bare feet in the water.

"'What are you doing? They stink, you know!' Violalibera giggled, but Cesare's solemnity quickly silenced her.

"He scrubbed their feet, one at a time, until they were clean. Their two pairs of feet side by side, as radiant as those of a bride and groom. Violalibera wriggled hers around in the basin, splashing a little. Then we all smiled. The tension dissolved like the dirt in the water. Once again there was someone to decide for us.

"Then Cesare dried them off with a dish towel. He knelt for so long that when it came time to get up, he had to cling to the tabletop.

"'I know what you had in mind,' he said, 'but it was fear that generated those thoughts. Now they're gone. This child will be born. Take each other's hands. That's right. Pray with me.'

"The doctor arrived half an hour later. He examined Violalibera in our room and found her malnourished. He prescribed absolute bed rest and some medicines. The following day, Cesare and Bern would have to take her to a specialist for an ultrasound. Nicola and I were still there, in the kitchen, but already like a pair of spectators.

"Since the lawn mowing awaited me at the Relais dei Saraceni in a few hours, I said goodbye. In the Atala's rearview mirror, the masseria became an ever-tinier spot, then it disappeared altogether."

TOMMASO'S VOICE was strained by exhaustion or maybe by those memories.

Throughout that long night in which he'd talked and I'd listened, in which he occupied half of the double bed and I a chair that felt more and more uncomfortable, during all that time our eyes had met only a few times. We chose to stare in turn at a piece of the bedspread, at the clothes spilling out of the open closet, at Medea's moist snout. But now I couldn't take my eyes off him, I couldn't stop wondering how he had managed to conceal it all behind that pallor of his, and for so long. And how had Bern been able to? But the words remained stuck in my throat. I drank the last sip of water in his glass to swallow them, along with the image of Bern and Violalibera's feet side by side in the basin and his silent consent to be the father of that baby. Their baby.

"It was Nicola who told me what happened next, by phone," Tommaso resumed. "One of the waitresses came to call me while I was picking string beans."

He sighed. "Violalibera had probably thought that a dozen leaves would be an adequate number, lethal for the baby but not for her. Nicola explained that she had sweetened the oleander tea before drinking it. Then she went out and walked as far as the reed bed. They'd found her several hours later, Yoan did. When the medics arrived, she was still breathing, she was still breathing in the hospital too, but by evening she was dead. When he heard, Bern fled to the tower, but Cesare and Floriana didn't go looking for him again.

"You arrived a few weeks later. The night Nicola took you to the Scalo, I was there with Bern. We were standing near the tower, at the darkest spot. You had your back to us, but there was a moment when you turned around. It felt as if you were looking right in our direction, right at us. I remember thinking: It's as if she caught the scent of something in the air. At that moment if we had simply moved a hand, you'd have noticed us. In fact, Bern took a step forward, toward the light, but I held him back. Hadn't we had enough troubles already? And when that moment had passed, you turned back to Nicola.

"In the fall, Cesare and Floriana also left the masseria. They went off without seeing to a thing; they packed the Ford's trunk and left. They didn't even secure the iron bar across the dirt track. As if that death had cursed the land forever, as if all the prayer Cesare was capable of could not suffice to purify it."

TOMMASO REMAINED SILENT for a few seconds, as if to allow me time to fully absorb that information as well.

Then he added: "She had tied her wrists with a rope. The same one that Bern and I had used to drag the trunk of the palm tree. So she wouldn't instinctively try to run away and call for help before the abortion was complete. I don't know where she had learned to make a knot like that, not everyone can. She had vomited on herself, bound like that. It seems that after drinking an extract of oleander cramping occurs right away, but the poison takes several hours to reach the heart. The heartbeat slows almost to a stop, then wildly accelerates again. Yoan told Nicola, and Nicola told me, that Violalibera's body was so light that the boy had picked her up effortlessly. He ran to the house, carrying her in his arms, and placed her on the swing-chair. When Floriana opened her eyelids, her eyes were all white. Bern was there, watching."

Tommaso took Stirner's book from the nightstand. He opened it.

"I read it just a short time ago. It's a terribly boring book. Boring and confused. Or maybe I'm not intelligent enough to understand it. Anyway, I found the words Bern had recited from memory in the olive grove, before we threw the stones."

He leafed through the pages until he found what he was looking for.

"'*The truth is dead, it's a letter, a word, a material that I can use up.*' That's what Bern said."

"It's just a phrase, like any other," I replied, but I struggled with it. I had a hard time speaking.

He put the book back on the bedside table, stared at it for another moment.

"All three of us kept the promise. We never spoke about Violalibera again, not with others or among ourselves. At least not until tonight."

PART TWO

THE

ENCAMPMENT

3.

I was twenty-three when my grandmother died. After the summer of my senior year of high school I had seen her only once: she had come to Turin for a medical exam, her throat, or maybe her ear, and had stayed two nights in a hotel. But one evening she had dinner with us, and she and my mother talked about this and that with the utmost cordiality. As she was leaving, she asked me if I'd enjoyed the book that she had sent me with my father. I barely remembered it, but I said yes so as not to offend her.

"Then I'll send you some others," she promised, though afterward she must have forgotten.

Nobody knew when she'd gotten into the habit of going to the beach in the morning.

"In February! Swimming in February!" my father ranted. "Do you have any idea how cold the water is in February?"

My mother stroked the sleeve of his jacket; he was shivering uncontrollably.

A fisherman had spotted the corpse slamming against the rocky shore at Cala dei Ginepri. I knew that cove, and all afternoon I kept seeing my grandmother's body knocking brutally against the rocks. When they'd pulled her out, she had been soaking for hours, the skin of her face and fingers was all wrinkled, the knees she was so ashamed of nibbled by the bottom fish.

My father decided to leave that same day. In the car no one spoke, so I dozed in the backseat. When we arrived in Speziale it was dawn and a layer of fog hung over the countryside.

I wandered, dazed, through the courtyard of the villa with an awful taste in my mouth. I approached the pool, which was covered by a canvas tarp: a calcareous ring gleamed in the center. I stepped on one of the waterlogged cushions surrounding the tile edge. Everything spoke of neglect.

The coming and going continued until suppertime. I recognized some of my grandmother's pupils, now teenagers yet escorted by their mothers. They spoke of her as "the teacher," and took turns sitting on the couch that had been her sanctuary as they offered their whispered condolences to my father.

The windows were wide open and gusts of cold air chilled the room. I didn't go near the open coffin in the middle of the room. Rosa offered the visitors small glasses of liqueur and marzipan confections. Cosimo was leaning against the wall with his hands clasped and a beaten look, as my mother spoke to him up close.

Suddenly she left him standing there and headed toward me.

"Come with me," she said, grabbing my arm.

She led me to my room, where nothing, absolutely nothing, had changed since the last summer I'd been there.

"Did you know there was a will?"

"What will?"

"Don't lie to me, Teresa. Don't even try. I know you had a special bond, you and she."

"But I never even phoned her."

"She left it all to you. The house. Along with the furniture and the land. Even the lodge where Cosimo and that insufferable wife of his live."

I didn't immediately understand the significance of what she was telling me. The will, Cosimo, the furniture. I was overcome by the unexpected emotion that the made-up bed had stirred in me.

"Listen to me, Teresa. You will sell this house right away, you will not listen to what your father tells you. It's nothing but a decrepit villa, full of leaks. Cosimo is prepared to buy it. Let me take care of it."

THE FUNERAL WAS HELD the following day. The church in Speziale was too small to hold everyone, and many people gathered in the doorway, blocking the light. At the end of the service the priest approached our pew and shook my hands.

"You must be Teresa. Your grandmother spoke a lot about you."

"Really?"

"Are you surprised?" he asked, smiling at me. Then he patted me on the shoulder.

We followed the coffin to the cemetery. A niche had been opened next to that of my grandfather and certain ancestors whom I knew nothing about. When the undertaker started fumbling with the trowel and the casket was lifted by the hoist, my father started sobbing. I looked away, and it was then that I saw him.

He stood a distance away, hidden behind a column. Bern. His clothes struck me as the most noticeable sign of how grown up we'd become. He wore a dark coat and, under it, a knotted tie. Meeting my gaze, he ran a forefinger over his eyebrow, and I didn't know if it was just an awkward gesture or a secret code that I no longer knew how to decipher. Then he moved quickly toward one of the family chapels and disappeared inside. When I looked back at the coffin, which was now being pushed into the niche, rasping and creaking, I was so confused, so distracted, that I didn't give my grandmother even one last thought.

As the crowd began to disperse, I murmured to my mother that I would join her later at the house, there were some people I wanted to greet. I walked around the churchyard, very slowly. By the time I reached the gate again, everyone had left. I went back into the cemetery. The undertaker, left to himself, was finishing the job of sealing up the marble slab. I looked in the chapel, but Bern wasn't there.

I returned to town, almost running. Instead of turning in to my grandmother's villa, I continued on to the masseria. The iron bar across the entrance was open. Going back down the dirt track to the house again was like sinking bodily into a childhood memory, a memory that had lingered there, intact, waiting for me. I recognized every single thing, every tree, every cleft of every single stone.

I spotted Bern sitting under the pergola, along with some others. I had a moment's hesitation, because not even then, seeing me, did he encourage me to join him. But soon enough, there I was among them. Bern, Tommaso, Corinne, Danco, Giuliana: the people with whom I would share the next few years of my life—by far the best years, and the unsuspected prelude to the worst.

Bern introduced me in a neutral tone, saying that I was the teacher's granddaughter, that I lived in Turin, and that I used to spend vacations there at one time. Nothing more. Nothing that would let anyone think we had once been intimate, he and I. However, he stood up to get me a chair. Tommaso murmured his condolences for my grandmother without looking at me directly. A wool hat covered his fair hair, and his cheeks were reddened by the cold; seeing him nervously jiggling his leg, I had the old feeling that he didn't want me there.

Beers appeared on the table and Giuliana poured pistachios out of a plastic bag. Everyone scooped up a handful.

"I'd heard that the masseria was for sale," I said to break the silence, "but not that you'd bought it."

"Bought? Is that what you told her, Bern?" Danco asked.

"I didn't tell her anything."

"I'm sorry to disappoint you, Teresa. We didn't buy it. None of us would have the money."

"Corinne would," Giuliana objected. "All it would take is a phone call to her daddy, right?"

Corinne raised her middle finger.

"So you're renting, then?"

This time they laughed heartily. Only Tommaso remained serious.

"I see you have a rather canonical idea of private property," Danco said.

Bern gave me a brief glance. He was sprawled in his chair, his hands in his coat pockets.

"I think you could call us squatters," he explained briefly, "even though Cesare probably knows we're here. But this place no longer interests him. He lives in Monopoli now."

"We're squatters, so we don't have electricity," Corinne said. "A real pain in the ass."

"We have the generator," Danco countered.

"Sure, we keep it on for an hour a day!"

"Thoreau lived beside a frozen lake with no electricity," he persisted. "Here the temperature never drops below freezing."

"Too bad Thoreau didn't have long hair down to his ass like me."

Corinne stood up to move closer to Tommaso; he pushed his chair back and made her sit on his lap. "I'm still chilly. Rub me hard," she told him, snuggling against his chest. "Not like a fucking kitten, vigorously!"

Scratching something off her sweater, Giuliana said: "Anyhow, a long cable is all we'd need to tap into Enel's power line."

"We've already discussed this," Danco replied, "and we voted, I believe. If they found out we're stealing electricity, they'd make us clear out. And after a while, they'd be sure to notice."

Corinne looked at him coldly. "Will you stop throwing the shells on the ground?"

"They're bio-de-gra-da-ble," Danco said with a defiant smile, as he tossed another pistachio shell over his shoulder.

I felt Giuliana staring at me, but I didn't dare turn around to face her. I slowly brought the bottle of beer to my lips, trying to overcome my uneasiness.

"So, what do you do in Turin?" she asked.

"I'm studying. At the university."

"And what are you studying?"

"Natural science. I hope to become a marine biologist."

Danco started sniggering. Corinne punched him in the chest with a fist hidden in the sleeve of her sweatshirt.

"Teresa, who once lived underwater," Tommaso remarked faintly.

Corinne rolled her eyes. "Oh, boy! Not that game again!"

"Do you like horses, too?" Danco asked. He was serious now.

"I like all animals."

I noticed an exchange of glances among them, but no one spoke. Then Danco said, "Excellent," as if I had just passed a test.

After a few minutes spent drinking in silence, with Corinne tormenting Tommaso, tickling him behind his ear, I asked: "And Nicola?"

Bern finished his beer in one swig and slammed the bottle on the table. "Living the good life in Bari."

"By now he must have graduated."

"He dropped out of university," he said, more and more sullenly. "He preferred to join the guards. They reflect his personality better, apparently."

"What guards?"

"The police." Giuliana stepped in. "What do you call the guards in Turin?"

Tommaso said, "It's been two years now."

"Left! Right! Left! Right!" Danco barked, swinging his arms rigidly.

"I don't think the police march," Corinne said.

Giuliana lit a cigarette and tossed the pack on the table.

"Another one?" Danco asked angrily.

"It's only the second one."

"Great. So it's only another ten years of synthetic waste," he kept on.

Giuliana took a long drag and blew the smoke toward him spitefully. Danco held her look impassively.

Then he turned to me. "Do you know how long it takes a cigarette to decompose? Something like ten years. The problem is the filter. Even if you crush it in the end, as Giuliana does, it doesn't change anything."

I asked her if I could take one.

"The first rule of the masseria," she said, pushing the pack toward the center of the table. "You never have to ask permission here."

"Forget your concept of ownership," Danco added.

"If you can," she finished.

Corinne said, "I'm hungry. And I warn you, I'm not going to lunch on pistachios again. Today it's your turn, Danco, get a move on."

Instead they started talking among themselves, as if they'd forgotten that I was there. I leaned toward Bern and in a low voice asked him if he wanted to walk me home. He thought about it for a moment before getting up. The others paid no attention to us as we walked away.

AND SO there we were, retracing the same route we'd walked as kids. The countryside in winter was different, more melancholy, I wasn't used to it. The soil, a dusty red in August, was covered with a mantle of tall, gleaming grass. Bern didn't speak, so I said: "Those clothes look good on you."

"They're Danco's. They're too big, though. See?"

He turned the cuff over: it had been folded inside and held with a safety pin to make the sleeve look shorter. I smiled.

"Why didn't you wait for me after the funeral?"

"It was better that they not see me."

"Who?"

But he didn't answer. He kept his eyes on the ground.

"There were so many people," I said. "I wouldn't have imagined it. Nonna was always alone."

"She was a generous person."

"And how would you know?"

Bern raised his collar, then lowered it again. Wearing that coat seemed to take up a lot of his energy.

"For a while she helped me study."

"My grandmother?"

He nodded, still looking down at the path.

"I don't understand."

"I wanted to take the exam to get into the fourth year, but in the end I dropped the idea."

He had quickened his step. He sighed. "In exchange for lessons, I helped Cosimo in the fields."

"And where did you live?"

"Here."

"Here?"

I felt light-headed, but Bern didn't notice.

"When I heard that Cesare and Floriana had gone, I decided to come back. I'd been staying at the Scalo for a while, in the tower. I brought you there once."

He'd been living there, at the masseria, exactly where I'd imagined him all that time, with Violalibera and their child. I must have been thinking something like that and no doubt I wondered where they were at that moment—"It seems that Bern got into some trouble . . ."—but I couldn't speak.

"I didn't know," I murmured instead. "Nonna didn't tell me."

Bern looked at me for a moment. "Really?"

I nodded. I felt weak.

"That's odd. I was sure you knew. I thought you weren't interested in coming anymore."

After a pause he added: "Maybe it was for the best. Better for you."

"Why the hell didn't she tell me!"

"Calm down."

But I couldn't, I became hysterical, I kept repeating why why why, until Bern gripped my shoulder.

"Calm down, Teresa. Sit here a second."

I sat down on a stone wall; I was breathing heavily. He waited patiently at my side. Then he bent down and broke off a leaf, rubbed it in his hands until it was crushed, and held it to my nose. "Smell this."

I inhaled deeply, but the scent I recognized wasn't the plant's, it was his skin.

"Mallow," he said, sniffing in turn. "It helps relax the nerves."

Sitting on the wall, we gazed at the green, hushed countryside for a while. I was calmer, but with the calmness had come an overwhelming weariness, together with a kind of regret.

"Did she go swimming even then?" I asked.

"I went with her sometimes. I sat on the sand. She would swim way out to sea, on her back; all I could see was the pink dot of her bathing cap. When she returned to shore I'd be waiting for her with an open towel and she would tell me, 'You don't know what you're missing.' She always said that."

I was suddenly hungry to look at him, to touch him, the desire that stirred in me was frightening. I slipped my arm under his and pressed against him.

"You're crushing me," he said.

I quickly drew my hands back into my jacket pockets. What right did I have to grab him that way?

"I didn't say stop. Only to loosen your grip."

But I kept my hands in my pockets. I stood up and started walking faster than before, as if fleeing from that moment of weakness. The landscape suddenly changed before us. We were at the edge of a grove of trees that were shorter than the olive trees and had white blossoms on their leafless branches.

"Here we are," Bern announced, as if his intention from the beginning had been to take me there.

"What are they?"

"Almond trees. I figured you'd never seen them like this. This year they bloomed early. And now the cold is threatening to ruin everything."

We walked into the orchard, the heels of my shoes sinking into the soft clods.

"I'll break off a branch for you if you want."

"No. They're best seen from here."

"Remember when you left me the Walkman among the almond shells? Sometimes in the tower I felt lonely, so I listened to your tape. I always listened to it from beginning to end, until it wore out."

"It was awful music."

Bern looked at me as if he didn't understand. "It was beautiful music."

After a few minutes, to my surprise, we found ourselves in front of the gate to the villa.

"When do you leave?"

"Today. Soon."

He nodded. I thought six hundred miles of highway would be enough. There was so much waiting for me in Turin: my university courses, the exams, more courses, and a thesis to choose. Everything would fall back in place. But just then Bern looked up and his close-set eyes had the same effect on me as when I was a girl, the first time our eyes met as we stood on opposite sides of the doorway to my grandmother's house. I'm almost certain it was I who leaned over and kissed him on the lips.

"Why?" he asked me candidly, after having kissed me back. He had a melancholy smile, which disturbed me even more. Why? Because I'd wanted nothing other since the day I went to look for him at the masseria and hadn't found him there, as if everything had been left on hold since then. The fact that at some point I had forgotten that desire didn't mean that it wasn't still there, vivid, unchanged.

But instead of confessing it to him I asked: "Do you have a child, Bern?"

He pulled back a little. He glanced away for a moment.

"No. I don't have a child."

"And that girl?"

I lacked the breath to say her name.

"There is no girl, Teresa."

I believed him. Every fiber in my body wanted to believe him. We would never talk about it again.

"WHERE WERE YOU?" my mother asked when I entered the house. "Your father wants to leave right away. He wasn't able to sleep, poor thing. I'll have to drive. Rosa made us some sandwiches, we'll eat in the car."

A number of objects had disappeared from the living room: the silver frames with the photographs inside, a vase, the clock supported by the trunks of two elephants. In an open bag beside the door I saw brass gleaming. My mother intercepted that look.

"Check if there's anything else you want to take."

I filled the overnight bag with the few clothes I had brought. From the window of the bedroom I watched my parents in the yard with

Cosimo and Rosa, the car doors already open. My father raised his eyes in my direction, probably without seeing me. I was having a little trouble breathing. I sat on the bed, next to the already closed suitcase, and stayed there motionless for a few minutes. In that very short time, I made up my mind without really deciding. Descending the stairs, I felt weightless, as though my feet were barely touching the steps.

"Your things?" my mother asked.

"They're upstairs."

"And you didn't bring them down? Wake up!"

"I'm staying here."

My father whirled around, but it was she who spoke again: "What are you talking about? Move, hurry up!"

"I'm staying. For a couple of nights. Rosa and Cosimo can look after me, can't you?" The caretakers nodded, somewhat incredulous.

"And what do you intend to do here, if I may ask?" My mother kept at it. "Cosimo has already turned off the heating."

Then my father said, "You saw him."

There was no trace of irritation in his voice, only an extraordinary weariness. My grandmother was dead and he hadn't slept for all those hours.

"Who are you talking about?" my mother persisted. "You're driving me crazy, Teresa. I'm warning you."

But I was no longer listening to her. She knew nothing about that place, she didn't understand and would never understand. My father did, however. Because the two of us had been infected with Speziale in the same way.

"Did you see him?"

I couldn't meet his eyes.

"Get in the car, Teresa."

"Just a couple of nights. I'll take the train back to Turin."

"We're leaving now!"

The caretakers watched us. My father put his hand on the car door. His eyelids were purplish.

"You knew," I said, almost in a whisper.

He turned to me. For a moment his eyes widened.

"You knew he was here and you didn't tell me."

"I didn't know a thing," he snapped, but there was a slight uncertainty in his voice.

"How could you?"

"Let's go, Mavi," he said then to my mother.

"You want to leave her here? Are you crazy?"

"Get in the car, I told you."

He shook hands with Rosa and Cosimo, murmured some instructions, then sat down in the driver's seat.

"I'll expect you at home. Two days at the most."

He started the car, then seemed to have an afterthought. He twisted in the seat to wrest his wallet out of his pants pocket, took out a few bills, and handed them to me without counting them.

A few seconds later they were gone and I was standing in the yard with the caretakers, surrounded by the silence of the countryside.

BETTER TO WAIT until tomorrow, I told myself, not go back right away, otherwise he'll think I postponed my departure because of him. But there was nothing in my grandmother's house that could detain me, only impatience, so two hours later I was back at the masseria.

They were all outside, gathered around a strange object, a kind of overturned umbrella clad in aluminum.

"Let's see if she at least can guess what it is," Danco said, not showing the slightest surprise at seeing me back there.

"A satellite dish?" I suggested.

"I told you so!" Corinne exclaimed. "It's impossible."

"Try again, come on," Danco prodded me.

"A giant frying pan?"

Giuliana made a scornful face.

"Getting warmer," Tommaso said.

Corinne lost her patience. "So tell her!"

"This is progress, Teresa. Innovation combined with respect for the environment. It's a parabolic solar concentrator. If you put an egg here in the middle, you can cook it using only sunlight. In the summer, of course."

"Too bad it's February," Corinne retorted.

Then she took advantage of my less-than-enthusiastic reaction to needle him some more: "You see? She thinks it's a load of crap. Danco bought it with our communal kitty without even asking us."

"I don't think it's a load of crap," I said uncertainly.

"Maybe we can still return it," Tommaso suggested.

"Just you try it!" Danco threatened.

Bern was watching me, but differently from the way he had that morning, as if he had suddenly remembered something.

"So you stayed," he said softly.

Danco announced that it was time to get back to work. He waved his arms to scatter us.

"Come and help me in the food forest?" Bern asked me, and I said yes, even though I didn't know what he was talking about.

"Weren't there oleanders here?" I asked as we walked off from the house.

"We let them dry up two summers ago," he said. "They took too much water. Cesare was incredibly irresponsible about these things. He thought not killing anything was enough to save us."

"Save us from what?"

Bern gave me a steady look. "By now the water supply is about to run out. Do you know what happens when you pump water from all the artesian wells there are here?" I didn't know, of course. "The aquifer is drained and filled by seawater. If we go on like this our land will become a desert. What we have to do is regenerate," he said, emphasizing that word "regenerate."

It occurred to me that without electricity they couldn't use the well anyway. Whenever the power failed at my grandmother's, nothing came out of the taps.

I asked him how they managed, and he turned around to face me, though he didn't stop walking.

"If you can't steal water from the ground, where do you get it?" he said, pointing up.

"You mean you do everything with rainwater?"

He nodded.

"And you drink it too? But isn't it full of germs?"

"We filter it with hemp. I'll show you later, if you want."

In the meantime, we had reached the mulberry tree. I found it hard to recognize it that way, completely bare. Vegetation had grown up all around it, which at first glance seemed out of control: saplings; artichoke, pumpkin, and cauliflower plants; and all kinds of weeds.

"It's better to work with our hands," Bern said, bending to the ground. "We have to clear all these away."

He grabbed a mushy handful of rotten leaves and tossed them behind him. "We'll pile them here. Then I'll come back with the wheelbarrow."

"Why did you let it go like this?" I asked, kneeling beside him a little reluctantly because those were the only jeans I had with me.

"Let what go?"

"The vegetable garden. It's a mess."

"You're wrong, everything is exactly in its place. Danco spent months planning the food forest."

"You mean you chose to plant the trees and everything else this way?"

"Don't stop clearing the leaves while you're talking," Bern said, glancing at my hands. He took a deep breath. "The mulberry ensures shade in summer. And we pruned it to make it expand as much as possible. Around it are fruit trees and under them the legumes, which serve to fix the nitrogen."

"You talk like an expert."

He shrugged. "It's all thanks to Danco."

The soil beneath the decaying leaves was warm. By then the knees of my jeans were stained, so I figured I might as well get comfortable. I scooped up ever bigger armfuls and threw them onto the pile.

"We're just about self-sufficient," Bern said, "and soon we'll be able to sell part of the harvest. Now everything you see is bare, but in summer the production is copious."

"Copious," I repeated slowly.

"Copious, yes. Why?"

"Nothing. I had just forgotten about the words you use sometimes."

He nodded, but as if he didn't fully understand.

"And why are we clearing away these leaves now?" I asked. I felt like laughing, I didn't know why.

"It's better to remove the mulch before spring arrives. So the warmth can penetrate the soil."

"Danco says so, I imagine."

I meant to provoke him with a joke, but he answered me with the utmost seriousness: "Yes. Danco says so."

We spent another half hour in almost complete silence. I was beginning to sense that Bern would not ask me anything about my life in the years we'd been apart, just as he didn't when he was a boy. As if what happened a distance away from him, far from the trunk of that mulberry tree, didn't exist at all, or in any case had no importance. But it was fine just the same. It was enough for me to be near him, to grope among the plants and breathe that moisture-laden air together.

I LINGERED at the masseria until sunset, then until supper, each time telling myself that I would leave right afterward. We ate a mishmash of eggs and zucchini cooked by Corinne, completely unsalted, though I didn't dare say anything because everyone seemed to like it. I was still hungry, but there was nothing else, so I kept nibbling on the bread; I had the impression that Giuliana was counting every mouthful.

The sole hour of electricity for the day ended with the meal, and we moved in front of the lit fireplace. In addition to that, there were only a few candles to illuminate the room, some half melted on the floor. Although we huddled as close as possible to one another and draped blankets over our shoulders, it was cold. Yet even then I didn't really consider the idea of going home, of leaving Bern and the others and the fire that gleamed in their eyes.

Around eight o'clock Danco shrugged off his blanket and said it was time to move. They all jumped up and for a few seconds I was the only one still sitting on the floor. Looking down at me, Danco said, "Are you coming with us?"

Before I had time to ask where, Giuliana started protesting that there wasn't enough room in the jeep. But he ignored her.

"You came on a special week, Teresa," he went on. "We have an action planned for tonight."

"What action?"

"We'll explain it to you in the car. You need black clothes."

Until a moment ago they were all numb, about to fall asleep, but now a wild excitement electrified them.

"All I have is my funeral outfit," I said, becoming more confused, "but it's at the villa."

"That's all we'd need, for you to come in that!" Giuliana exclaimed. "Stay here, Teresa. Believe me, you're better off."

She patted my cheek, but Danco silenced her once and for all: "Stop it, Giuli. We already talked about it."

Corinne took me by the arm. "Come on. We have a full closet upstairs."

We went up, we three girls, and Corinne began rummaging through a pile of garments scrunched up like rags, while Giuliana undressed.

"Whose are they?" I asked.

"Ours. I mean, everyone's. This is the girls' side."

"You keep your clothes mixed together?"

Giuliana laughed and said caustically, "Yeah, that's right, mixed together. But don't worry, they're clean."

Meanwhile, Corinne had extracted some black leggings. "Try these," she said, throwing them at me. "And this," she added, digging out a sweatshirt similar to her own.

She didn't look away as I took off my sweater.

"You have fabulous tits. See them, Giuli? You'd only need a fourth of hers not to look like a man."

I didn't dare object that the leggings didn't fit me well, that according to my mother I didn't have the right body to wear anything close-fitting, or that I would probably freeze to death.

"Stop looking at yourself," Giuliana said, "we're not going to a fashion show."

Four of us were squeezed into the backseat of the jeep: we girls and Tommaso, who stared obstinately at the dark fields beyond the edge of the highway.

"Where are we going?" I asked.

"Foggia," Danco replied.

"But it will take at least three hours!"

"Roughly," he said, his tone expressionless. "You should get some sleep."

But I didn't feel like sleeping. I persisted with my questions, and finally Danco decided to explain to me what the "action," as they called it, consisted of. He spoke in a low voice, forcing me to stay focused. He said there was a slaughterhouse in San Severo, that horses were brought there from all over Europe after being made to trudge thousands of miles without food or water. And that the methods used to slaughter them were brutal.

"A gunshot to the neck before butchering them," he spelled out. "It might seem like a quick death, but those awaiting their turn see everything that's happening and they start struggling, so they beat them with clubs to stun them. That's where we're going, Teresa. To a nightmare."

"And once we get there, what do we do?"

Danco smiled at me in the rearview mirror. "We free the horses, right?"

WE REACHED the slaughterhouse late at night. Tension had kept me awake listening to the jazz music that drifted from the car radio. I was no longer so sure that following them there was a good idea.

We left the jeep in a place sheltered by trees, then continued on foot along the edge of a field. There was a little bit of moonlight, just enough not to stumble.

"What if they spot us?" I asked Bern in a whisper.

"It's never happened."

"But what if it happens?"

"It won't happen."

The barn stood out in the distance; a floodlight illuminated the yard in front.

"They're in there." Danco pointed. Surprisingly, Bern put a hand on my neck. "You're shivering," he said.

Forcing the padlock on the gate was easy. We inched forward, hugging the surrounding wall. I could feel the night's dampness through the thin fabric of the leggings. For a moment I saw myself through the eyes of those who knew me in Turin. What the hell was I doing there? But the hesitation gave way to unbridled glee.

Giuliana and I were ordered to keep an eye on the owners' house. The windows were all dark.

"So you and Bern were together?" she asked me as soon as we were alone.

"Yes," I replied, even though I wasn't sure it was true.

"And how long has it been since you've seen him?"

"A long time."

We could hear the others struggling with the cutters behind us, cursing because the lock was more resistant than the one at the gate.

"Are you and Danco together?"

Giuliana raised her eyebrows. "Sometimes."

Then a different, sharp clack was heard, followed by the sound of a chain dropping onto the concrete. We turned at the exact moment the door opened and an alarm began to shriek.

Lights immediately went on in the house, one, two, three of them. Bern and the others had disappeared.

"Come on, fuck!" Giuliana yelled, pulling me by the arm.

I found myself inside the barn, in semidarkness. Danco and Bern and Tommaso and Corinne were opening the doors of the stalls and shouting at the horses to get out, slapping their flanks. As if roused, I started doing it too, but the horses wouldn't move, they just stamped their hoofs, agitated by the sound of the siren.

"They're coming!" Corinne hollered.

Then Tommaso did something. "I nipped a horse with the cutters," he would explain to us later, in the car, as we drove back, careening along the highway, charged with adrenaline, everyone talking over one another.

The horse that was nipped started galloping toward the way out and all hell broke loose. The others followed him, hurtling into one another. I stood flattened against a column so I wouldn't get crushed, until Bern appeared beside me, out of nowhere, emerging from the dynamic throng of manes and hoofs.

We ran out behind the last of the animals. There were men in the yard, but they couldn't decide whether to stop the horses or us, so we gained an advantage. We fled through the field. I could see Danco ahead of everyone, and Corinne.

There were shots. The horses got even more frantic, but they were running around in circles. They had scattered in front of the barn, because there hadn't been time to push them outside the wall; only a few of them had gotten the idea.

The men had given up pursuing us, one of them was closing the gate, another was chasing after the runaway animals. We enjoyed that sight of freedom for a few seconds.

"We did it, shit!" Tommaso shouted. I had never seen him like that before.

On the way back, when the enthusiasm had partially subsided, some of us closed our eyes, and sweaty heads slumped onto the shoulder of whoever was beside them; mine on Bern's shoulder, and he didn't move a muscle after that so as not to wake me up.

I dreamed that the liberated horses were a whole herd, running through a barren clearing, raising a cloud of dust so thick that they seemed to be floating in air. They were all black. I didn't merely watch them, and I wasn't one of them, I was more than that: I was that entire multitude.

IN THE MORNING I was awakened by a hand caressing my face. A residue of the night's electric charge still lingered in the air. I had a rather confused memory of when we had all been together, drinking wine in the kitchen. Tommaso and Corinne had been the first to go, then Danco and Giuliana, or maybe the other way around. In any case,

Bern and I had found ourselves alone, and our dazed state had driven us up the stairs, to his room, into his icy bed.

But I remembered exactly what had happened next, what he'd done to me and I to him, the passion with which he'd taken me and the excitement, so intense that I hurt all over. And then when he'd reached for me a second time, but slowly, almost methodically. We'd repeated every one of the secret gestures from the reed bed; the memory of our bodies was breathtaking.

Now he smoothed the hair from my forehead, parting it in the middle as if trying to re-create the style I'd worn during our last summer together.

"The others are already downstairs," he said.

I was so sleepy I could barely talk. And the strange taste I had in my mouth embarrassed me, maybe he tasted it too.

"What time is it?"

"Seven o'clock. We get started as soon as it's light here," he said, tucking a strand behind my ear and smiling, as if he had finally found what he wanted. "The water to wash up is cold, I'm sorry. I can boil some in a pot."

I studied him intently. It was heart-wrenching to have him so close.

"I have to leave," I said.

Bern slipped out of the covers and, naked as he was, stood in front of the window, his back to me. He was still worrisomely thin.

"What are you waiting for, then? Get dressed."

"You'll catch cold. Come back under the covers."

"I hope you enjoyed the diversion." He grabbed his clothes piled on the floor and left the room carrying them under his arm.

A few minutes later I heard his voice alternating with that of Giuliana. I groped around for the phone on the nightstand. I had turned it off the afternoon before to conserve the battery. The signal was weak, but it was enough to enable the screen to fill with notices, about a dozen messages, all from my father. In the first he simply asked where I was, then he became more and more worried and finally furious.

Panicked, I typed in a response: "Sorry, battery was dead, I'll stay here until tomorrow, then I'll come home, promise." I sent it and a moment later the phone went dead.

Again I was greeted by the others with no surprise, as if I now lived there. The house was even colder than it was the night before, though the fireplace was lit. Corinne handed me a cup of coffee. I recognized Floriana's dishware.

"Well, Teresa has finally appeared, maybe she can serve as arbiter," Danco said.

"I doubt it," Tommaso said through clenched teeth.

"Tommi claims that today is not a good day to plant the chicory because, he says, we're in a waxing moon. I tried to explain to him that there isn't a single scientific reason why the moon should have anything to do with agriculture."

"For millennia farmers have waited for the waning moon to sow chicory," Tommaso broke in, "millennia. And you think you know better?"

"There it is! I knew it! I was sure it would come out sooner or later. Tradition." Danco stood up, all excited. "Until a few decades ago, people around here poured oil over their heads in the name of tradition, to ward off the evil eye. In the name of tradition, men have done nothing but slaughter one another."

He looked at Tommaso and me in turn.

"I'm glad you think it's funny," he said to me. "I'd laugh too, if it weren't the tenth time at least that we're having this discussion."

"Call it 'practice,' if you like that better than 'tradition,'" Tommaso shot back.

"Listen to me. First of all, none of the well-intentioned farmers you mention has a degree in physics like myself."

"You don't have a degree!" Corinne put in.

"All I'm lacking is the thesis."

"I know quite a few people who only lack the thesis."

"Secondly," Danco went on, raising his voice, "I'm still waiting for you to bring me a shred of scientific grounds. But fortunately, we now have Teresa, right? Maybe in the natural sciences they taught her something about the moon that they forgot to explain to me."

I shrugged. I didn't think he really expected an answer; it seemed like just a game they wanted to include me in. I held the coffee's steaming vapor under my chin.

"Well?" he pressed me.

Tommaso was staring at me, as if he were remembering something.

"If I'm not mistaken, they say that moonlight's power to penetrate the soil is greater than that of the sun," I said. "And that helps germination. But I'm not sure."

"Aha!" Tommaso leaped to his feet, pointing a finger at his opponent.

Danco started writhing in his chair, as though having a convulsion.

"A greater power to penetrate? And what the hell would that power of penetration be? I feel like I've landed in a coven of witch doctors, fuck! If we go on like this, we'll start doing a rain dance. Teresa, I had faith in your coming here. Finally an ally, I told myself. And now you're defending the phases of the moon. Power of penetration!"

"That seems to be just what interests her," Giuliana commented, causing a sudden silence.

I thought I'd die of embarrassment; I didn't dare look at Bern or anyone else. Then she herself said: "What? Can't we even joke?"

After breakfast we helped Tommaso plant the chicory seeds in the greenhouse. The technique seemed strange and ineffective to me: we formed little balls of clay with our fingers, then dropped them somewhat randomly into the pots.

"To imitate the wind," Bern explained gravely.

He no longer seemed angry, only sad.

Finally Tommaso slapped his hands on his pants and said: "It won't grow. Next time you'll listen to me."

HE WAS WRONG. The chicory grew and I was still at the masseria when it was ready to be transplanted to the food forest; and I was still there when it exploded in big, plump heads in early summer. At our last phone call, my father had sworn not to speak to me again until I returned home.

Except for him, I didn't miss anything from my life in Turin, but I didn't try to explain it to my mother or to anyone else who called to ask the reason for my disappearance: they wouldn't have understood. All that mattered was going to bed with Bern in the evening, finding him beside me in the morning, watching his eyelids still heavy with sleep, in a room that was only his and mine, from which all you could see were trees and sky. And the sex, especially the sex, blind, dizzying: in the first few months it gripped us like a fever.

But there was also the euphoria of finally having some real friends— no, more than that, brothers and sisters. It must have taken me some time, of course, to get used to the no-flush outdoor toilet, to the lack of privacy that it involved, as well as to the rationed electrical power, the rotten-tasting water, and the assigned work shifts: for cleaning, cooking, burning the garbage. But I can't remember those troublesome aspects. What I remember instead are the long interludes when we sat under the pergola, drinking beer and playing cards.

In any case, ours was "do-nothing" farming: not doing what nature could do for itself. We wanted to understand nature's intelligence and take full advantage of it. And we wanted to regenerate—to regenerate everything that had been brutally consumed on that land.

Danco guided us and in the meantime he studied us one by one. Once he made a very complicated analysis of Tommaso's personality, starting from the habit he had of always opening a new jar of jam before the old one was finished. I understood little or nothing of it, but I saw that Tommaso was upset. Corinne interceded to defend him: "Now you're even watching what we do with the jars? You're really twisted."

Putting together the fragments of the story, I reconstructed how Danco had come to the masseria along with Giuliana.

Bern had lived there by himself for about a year, the period when he had occasionally worked for my grandmother in exchange for her private lessons. Then Tommaso had decided to join him, and Corinne was already with him.

"It was hard," they said of those months. "Fortunately, Danco and Giuli arrived."

Bern and the others had met them at the market center in Brindisi, where they went to do the shopping because things cost less in the discount stores. There were several different versions of that afternoon, each of them had one, and during the first months at the masseria I heard all of them. The meeting in the parking lot of the supermarket had become legendary, and to some degree it became so for me as well.

"Love at first sight" was how Giuliana described it.

She, Danco, and some other people they named in passing were picketing at the entrance to the supermarket. They had stopped Bern at the exit. "Can I see what you've got there?" Danco had asked him.

Tommaso and Corinne were ready to leave, but Bern had already let him come over and obediently opened his bag. Rummaging through it, Danco had asked him, "Why do you buy this stuff? You seem like a decent guy. May I ask what you do?"

"I work on myself," Bern replied.

"And besides that?"

"Myself, period."

That answer had astonished Danco. He'd started explaining to Bern why the cheese he had in his plastic bag was poison, why the bag itself was an abomination, and how those tomatoes grown in Morocco, thousands of miles away, would bring the whole planet to ruin.

"They were just some fucking tomatoes!" Corinne always intruded at that point, a little worked up.

"I have a proposal," Danco had said to Bern. "If I've piqued your curiosity, come by here tomorrow, even if it's just to tell me that I'm talking bullshit and that you'd rather go on working only on yourself. If you come, I'll bring you something."

That evening, at the masseria, Bern hadn't touched any of the food. When he returned to Brindisi, Danco was waiting for him alone in the parking lot. He had brought him a copy of *The One-Straw Revolution*; not his own copy, one he'd purposely bought that morning.

Later they'd met again and Bern had invited Danco to the masseria. Danco still didn't have a specific project in mind, but he was looking for one. He was in touch with people who'd devoted themselves to new forms of agriculture, mostly other university dropouts like himself. When he'd seen Tommaso's little vegetable garden, he'd had a vision. That's how it began.

Much later, I too arrived.

And now, almost every evening, Danco still read that same book to us: "This straw appears small and light, and . . . could become powerful enough to move the country and the world."

When the book ended, we implored him to start over again. We especially liked the first few chapters, where Fukuoka discovers his mission after a night of sudden illumination, but we wanted Danco to skip the boring part about cultivating rice, because who would ever cultivate rice in Puglia? But not Danco, he insisted on reading everything, claiming we'd miss some important insights otherwise. Actually, he just wanted to test how faithful we were to the cause.

When we got to the part about the Four Principles, we recited them in chorus, joking about our dedication, yet believing in it with all our hearts: "No cultivation! No chemical fertilizer! No weedkilling! No dependence on artificial substances!"

We felt as if we were the start of something, the start of change. Every moment had the lucidity of an awakening.

We carried out two other "actions." For the first we stationed ourselves at one of the many illegal dumps, at night, cloaked in dark sheets, scaring to death anyone who approached with their garbage. But what aroused our outrage the most were the lawns: those well-groomed lawns in front of vacation rental homes, so perfect, so incongruous. At the masseria we saved every drop of water; even in the torrid days of June our vegetables had to grow with only the soil's moisture. We let them wither and sometimes die of thirst because that was the right thing to do, whereas that ornamental grass was abundantly irrigated with groundwater.

We'd kept an eye on a time-share in Carovigno for days and we knew that there were no longer any renters, only a farmer who came by a couple times a week to make sure everything was in order. The utility room wasn't even locked. Destroying the irrigation system's control unit seemed too violent, even though Giuliana insisted, so Danco started meticulously taking it apart with a screwdriver. We removed the motherboard and only smashed that, then we put the lid back on. In the end the unit looked identical to the way it had when we'd arrived.

Two days later we returned to check. The lawn had yellowed; in another forty-eight hours it would have withered completely. But the farmer must have noticed the problem and fixed it promptly, because the next time we went by the sprinklers were going full blast. The grass had already regained its color.

As the weeks passed, however, the actions became less frequent. Maybe because we hadn't racked up any great successes, or maybe because we were more and more involved in our project at the masseria, and less and less interested in what happened outside.

Giuliana obtained some Super Skunk seeds and Tommaso planted them behind the house, surrounded by citronella bushes. It grew extraordinarily well, bulging with sticky flowers that we let dry in the shade before mixing them with tobacco. Giuliana managed to raise a little money, selling it to an acquaintance in Brindisi. But we never went overboard; getting rich was not our plan.

"We don't need more money, we need more knowledge," Danco used to say.

Still, money was a nagging worry. The more we scorned it, the more time we spent talking about it. Just when we reduced our needs even further by voting on a more inferior brand of beer, the jeep's battery failed for the second time in a few months.

"Because it's an old piece of crap!" Corinne said.

"Watch what you say," Danco shot back, "this Willy's made it through World War Two."

Then, just a week after replacing the battery, the bridgework on Giuliana's bicuspids cracked and she had to find a dentist willing to be paid in installments.

The only one who had a steady job was Tommaso. Every morning he left for the Relais dei Saraceni on his motorbike, often returning late in the evening. There were times when he was so exhausted that he preferred to stay there to sleep. He made his entire salary available to us, handing it to Danco the same day he got paid. I never heard him complain.

AUGUST. Mounds of dried seaweed littered the beach of Torre Guaceto, tiny crabs popped out of the sand and then vanished. We had snuck into one of the little coves prohibited to tourists, because we weren't at all tourists, and because we didn't like restrictions.

Danco suggested an exercise: "We'll take turns undressing in front of the others. Not all together, though, that would be too easy. One at a time."

"If you think I'm going to undress in front of you, you can forget it!" Corinne exclaimed.

Danco replied patiently, "What do you think your bathing suit is hiding? Something mysterious? We can all imagine what's underneath. Anatomy, that's all."

"Good for you, keep on imagining it, then."

"It's just the perception you have of your body, Corinne. They taught you to think that there is something absolutely private under those square inches of synthetic fabric. It's a sign of your mental limitation. But there is nothing absolutely private."

"Cut it out, Danco! You just want to see my tits."

"No. What I'd like is for you to be free from prejudices. For all of you to be," he said, as he slid his own swim shorts down to his ankles. He stood naked in front of us, his back to the light, long enough for us to become familiar with the reddish hairs around his sex.

"Look at me, Corinne," he urged, "all of you, look at me, go on. I have nothing to hide from you. If I could slit open my stomach and show you my guts, I'd do it."

So we imitated him, one at a time, the males first, then us girls. My fingers were trembling as I fumbled for the hook behind my back; Bern came to my rescue. Finally, our bathing suits lay scattered over the mantle of seaweed, like scraps of an old skin.

But instead of disappearing, our embarrassment grew greater and greater as the minutes passed. Finally we dived into the turquoise water.

"Let's run naked on the big beach!" Giuliana said, exhilarated.

"They'll call the police."

"If we run, nothing will happen to us," Danco said.

"Together, though—don't leave anyone behind."

We grabbed our swimsuits and scrambled up the slope, then, like a bunch of wild natives, descended on the long stretch of beach dotted with umbrellas. I didn't think I'd have enough breath in me to make it to the far end.

The beachgoers raised up on their elbows to get a better look at us, the kids giggled, shocked, and there were even some approving whistles. Everyone ran so fast, Corinne and Giuliana in the lead, graceful as ostriches. When I fell behind the others, I heard a man comment as I passed, not seeing his face. His words would return to me many months later, when everything was falling apart: "Fools," he said. "Who knows what they think they're proving."

IN SEPTEMBER Cosimo showed up at the masseria. From the tractor he unloaded two jerry cans filled with a clear liquid. Bern invited him to sit down, offered him some wine, which Cosimo refused with a wave of his hand. "I brought you some dimethoate," he said. "With a summer like this the flies will come in droves. Your olives next to the fence already have holes in them."

"That's very kind of you," Danco said, getting up, "but you can take the cans back. We don't need them."

Cosimo looked bewildered. "You've already treated them?"

Danco crossed his arms. "No, sir. We haven't sprayed our olive trees with dimethoate. We prefer to avoid insecticides here. As well as herbicides and phytopharmaceuticals of any kind."

"But if you don't use dimethoate the flies will ruin all the olives. And then it will come to my place, too. You can't taste it in the oil."

Unable to completely hide his timidity, he added: "Everyone uses it."

Bern must have sensed my discomfort, because he hurried over to

Cosimo, lifted the cans by the handles, and said, "It was considerate of you, thank you."

But in a flash Danco's order came from behind him: "Leave them where they are, Bern. I don't want that nasty stuff coming into our house."

Bern sought his friend's eyes as if to say, It's just to be polite, it doesn't cost us anything, we'll bring them in and then we won't use them, but Danco was adamant. So Bern backed away, murmuring: "Thank you anyway."

We had mortified the man. Cosimo, a white-haired, thick-skinned farmer, humiliated by a group of presumptuous kids. Corinne was studiously removing something from under her fingernails. Giuliana jiggled the flint of her lighter, sending tiny sparks spilling out of her clenched fist.

"Wait, I'll help you," Bern said, bending down to the cans again. But this time it was Cosimo who stopped him brusquely.

"I can do it on my own."

After replacing the cans on the tractor, he put it into reverse and, kicking up a little mud from under the wheels, drove back down the dirt track. But not before throwing me a look full of reproach.

"There was no need to treat him like that," I said when he had driven off. We could still hear the jolting rumble of the tractor.

"Do you really want to season your salad with that stuff?" Danco asked. "The best organological quality it has is being carcinogenic. Let him pour the dimethoate down his well! Let him and his wife drink it!"

"He was just trying to help us."

"So try again, Cosimo, and maybe you'll have better luck," Danco said cheerfully.

He expected us to go along with him, but Giuliana was the only one who did. Danco turned serious again: "They'd use DDT if they could still find it at the supermarket. They spread their chemical crap all over the place. And they don't even know what the hell is in it. Did you see his face when I said 'phytopharmaceuticals'? He didn't even know the word!"

"What do we do about the flies?" Tommaso asked. He had walked to the nearest olive tree, picked a bunch of olives that were still small, and spilled them onto the table. "The larva is inside."

Danco fingered the olives. "Honey and vinegar in solution, ratio one to ten. It's been done for years in organic farming. The flies are attracted by the honey and the vinegar kills them. In a word, traps."

We got to work that afternoon. We filled about fifty plastic bottles, and hung them at different heights at the ends of the branches. When the work was done, the slanting light of sunset lit up the cylinders, making them seem like so many lanterns.

After supper Danco made us hurry and clear the table. On it he placed a square of cardboard and a can of paint left over from our recent projects.

"You write it," he said, offering me a brush: "'The Masseria. A Toxin-Free Zone.'"

The sign was affixed with wire to the center of the iron bar that marked the entrance to the dirt track, replacing the one that read FOR SALE. It would remain there for years, slowly faded by the sun and rain, a little less legible with each passing season, a little more discordant, a little more false.

THE TRAPS FILLED with flies. We emptied the bottles and refilled them several times during the autumn. The oil was abundant. Once we were finished on our own land, we worked for others to harvest their olives. We stationed ourselves in the piazza and beat the competition from professional cooperatives by offering rock-bottom rates, half of what they asked for. We drove as far north as Monopoli and south past Mesagne. Danco got hold of a tow truck from some old friends, and

Tommaso was able to get Cesare's mechanical defoliator working again. We must have looked bizarre and scruffy when we showed up at a place, at seven o'clock in the morning. You could read the same thought in the landowners' eyes: Where the hell did these people come from? But we were young and extraordinarily energetic, and we worked well together; at the end of the day they often gave us an extra tip.

If it wasn't raining, at lunch we would sit under one of the trees to eat sandwiches brought from home. If the owner wasn't around, Giuliana would pull out a joint, and when it was time to get back to work we felt light-headed and stupid, we couldn't stop laughing. Danco calculated that by the end of the season we would have picked at least a hundred tons of olives.

With the money we earned (not as much as we'd hoped for, all in all), we bought used hives and bees to populate them. After exhausting discussions, we decided to situate them near the reed bed, because that spot was far enough away from the house, protected from the tramontana, and because we could take advantage of the natural spring to grow flowers. But the first generation of bees died after less than a week. Driven by an old reflex, Tommaso and Bern dug a grave and poured the striped cadavers into it, under Danco's icy stare. No prayer was said, however, there were only more debates, even more heated, on what we had done wrong.

Finally, Bern obtained a manual of sustainable beekeeping from the library in Ostuni. I was assigned to study it and then instruct the others on how to manage the breeding. It worked. Danco did not fail to remark on it every single day, when he happily sank his spoon into the jar of dark honey. For a while Giuliana sarcastically called me the bees' "fairy godmother."

In February we celebrated the anniversary of my arrival. The day I'd moved there, scraping the plastic wheels of the suitcase along the dirt track, had been designated as founding day. While Danco made a heartfelt speech, I could hardly believe it had already been a year.

That evening we drank a lot and at one point Bern allowed himself to reveal a secret. He told about when he slept alone in the tower at the Scalo, how on certain nights the sea was so thunderous that it kept him awake. So he would put on the headphones of the Walkman I had given him, the volume turned up to maximum, and he'd feel safe again.

Don't tell them about it, I begged him silently as he spoke, keep at least this secret just for us. But he didn't stop, since even private ownership of memories was abolished at the masseria.

"I wore out every millimeter of that tape," he said, his words slurred and his lips stained dark by the wine.

"What tape?" Danco asked, somewhat uncertain. He didn't like it when someone else drew attention to himself for so long.

"A cassette with a lot of different singers. I never knew what it was called. What was the name of it, Teresa?"

"I don't know," I lied, "it was just a mixtape."

Bern didn't give up, he was flooded with emotion: "There was one I liked more than the others. I would listen to it, then rewind the tape to the beginning and listen to it again. I got to know the exact number of seconds I needed to hold the key down to rewind it."

With his eyes half closed, and an unguarded bliss on his face, he began singing the melody. I hadn't heard him sing since the early summers at the masseria and I wished he would go on, but Corinne jumped up: "I know it! It's by that girl. What's her name? Come on, Teresa, help me out!"

"I don't remember."

Danco burst into one of his vicious laughs. "Sure, of course, the redhead on the piano!"

I felt Tommaso watching me as I looked at Bern, still silently begging him, but now hoping for him to say something, to make them stop before they ruined everything.

But he remained silent, unable to even return my look. And when Danco said, "What a pathetic story!" I saw Bern swallow, then give his new brother, his new supreme guide, an embarrassed smile full of submissiveness.

IN THE SPRING I returned to Turin; it was the only time I did. Bern was opposed to the trip, but I had to go, I hadn't seen my parents in far too long. When he realized that he would not stop me, he cautioned me: "Don't let them persuade you to stay. I'll be counting the hours and minutes."

On the train, my fear grew. I arrived in Turin certain that my father would use force, that he would beat me up, then lock me in the house, segregated like an addict, the same brutal methods that Corinne's parents had employed with her. So unaccustomed to people by then, I proceeded down the platform and through the echoing halls of Porta Nuova station; my legs gave way at the thought of meeting him.

But instead he wasn't there. He simply hadn't come. My mother said he'd preferred not to.

"What did you expect, Teresa, a welcome-home party?"

We had lunch, the two of us by ourselves; it felt very strange. I looked at the cookie tin behind her, which had been on that shelf forever, and which most likely still held the Doria Bucaneve, the flower-shaped cookies with the hole in the middle. My father used to stack them on his pinkie fingers, three on each side, and then nibble them with a twisted face that made me laugh as a child.

A couple of times I tried to start a conversation about the masseria. I would have liked to tell my mother that we had bought some chickens and that now there were fresh eggs every morning. The next time, maybe, I would bring her a few, and with them some of our mulberry jam. I wanted her to know that we had saved enough money to buy solar panels: starting next week we would have electricity at all hours of the day, clean, free energy, as much as we wanted. I really wished I could tell her about it, just as I wished I could confide in her how discouraged Danco's words sometimes made me feel, as if I were shallow and had no opinions.

And I wanted to tell her about Bern, especially about him; she would fall in love with him if she listened to me attentively for once, and then she would convince my father to stop his absurd retaliation of silence. The entire situation that now appeared so weird to him would come to seem natural, as it was natural to me. But none of this came out of my mouth. I ate quickly, then retreated to my room.

My room: cozy and so childlike. Photographs on the wall that no longer spoke to me, my university books still piled on the desk. Could I have left them like that? Or was it just another of my parents' tacit messages? The whole house was strewn with emotional traps: honey to attract flies, vinegar to kill them.

I allowed myself a long bath, even though I was bothered by Danco's voice accusing me of wasting water. He spoke more and more often in my head, like a new conscience, severe and implacable. But the water was inviting and scented with lavender, and my body melted softly in its warmth. I surrendered to it.

Still barefoot and with my hair wrapped in a towel, I took a book from the shelf: the one by Martha Grimes that my grandmother had sent me years ago through my father. I sat on the floor, my back against the clothes closet, and riffled through the pages, forward and backward. In the middle, I found a Post-it. I recognized my grandmother's handwriting, the same script used to pen comments in the margins of her pupils' exercises:

Dear Teresa, I've thought about it a lot. You were right that day. When I was speaking with you by the pool, I confused the word "unhappiness" with its opposite.

The message continued on the back:

In my life I have seen so many people make the same mistake. And I don't want it to happen to you too, not through any fault of mine, at least. I saw your Bern at the masseria. I thought you should know it. My lips are sealed, though. Affectionately, Nonna.

I cried a little after reading it, mostly out of anger. Why hadn't she chosen a simpler way to communicate with me? Had she read so many of her thrillers that she thought she was one of those characters? But I also cried out of unexpected, overwhelming relief, because my grandmother had not betrayed me, and because those words, though discovered so late, were her blessing for the life that I had chosen.

At that moment it seemed absurd for me to be at home, I was a different person from the one who had grown up there. I had to go back to the masseria as soon as possible.

I asked my mother for the biggest suitcase she had, promising to get it back to her. "I'll ship it," I added, so she wouldn't delude herself about my returning.

I filled it with clothes that wouldn't humiliate me in front of Corinne and the others. The next day I was back on the train, with new confidence. By then I was part of Speziale. Only my phantom self had left the masseria to travel north. And it didn't matter that I hadn't seen my father; he had wanted it that way. I tried to distract myself by reading Nonna's novel, but I was too preoccupied. In the end I gave up and just gazed out the window, until it got dark.

FINALLY WE HAD ELECTRICITY. We had a chicken coop on wheels to move the hens to where they would fertilize the soil. We had vegetables all year round. We had a solar skillet to cook scrambled eggs and even tiny ceramic cylinders to purify rainwater better, a Japanese invention that Danco turned up.

Yet something was wrong below the surface. Giuliana and I barely spoke to each other. After more than a year, she still treated me like an intruder. And many of us—that is, all of us except Bern—were starting to question Danco's role of group leader.

But the ones who showed the most disturbing signs were Corinne and Tommaso. They lived in an alternating state between anger and morbid devotion. More and more frequently Tommaso would spend the night at the Relais dei Saraceni and Corinne would refuse to join us for supper. She'd lock herself in their room until morning, alone, without eating.

One day, when it was already late August, she took me by surprise. We were washing the cups from breakfast.

"How many times do you do it, you and Bern?" she asked out of the blue. I understood, but I bought some time.

"What?"

"More than once a week? Or less?" She stubbornly kept her eyes down, looking at the stacked-up cups.

"About that," I said.

"About what? Once a week?"

Much more often, I was about to reply, but I sensed that would have upset her.

"Yeah."

Corinne turned abruptly, grabbed the teaspoons from the table in a single bunch, and slammed them onto the cups.

So I offered: "Tommaso works a lot."

"What, are you trying to console me? Who the fuck do you think you are?"

She was gripping the edge of the sink with both hands.

"Anyway, you could make less noise, you two. It's disgusting!"

She opened the tap all the way, but quickly turned it off.

"That bitch Giuliana! Let her wash her own cup. I've told her over and over again not to stub out her cigarettes in them. They all suck in this place!"

ANOTHER TIME we were gathered under the pergola for breakfast; only Tommaso was missing. We heard the screams, three of them, close together.

Bern was the first to leap to his feet. He ran behind the house, through the olive grove, like a fury. He had a precise destination in mind, as if he knew exactly what had happened, as if he had seen it. Danco had immediately rushed after him, with me behind him. For an instant Corinne was paralyzed, then she too got up and ran after us.

Giuliana, however, didn't move until we reappeared carrying Tommaso's disfigured body. Corinne was weeping hysterically, Bern was still wearing the beekeeper's paper coverall, white from head to toe.

We had found Tommaso on his knees, with swarms of bees around his head, swirling and buzzing; he was trying to drive them off, waving his arms, before he collapsed to the ground, unconscious. He was wearing a short-sleeved red-and-blue-checked shirt, unbuttoned to the navel. The bees wouldn't let him alone, they were disoriented, as if incredulous at having brought down such an enormous animal.

Bern had kept us from getting close. He'd gone to the tool shed, still running, and when he came back he was wearing the paper coverall. With his hands he swept away the bees glued to Tommaso's hair, his clothes, and the rest of his body. Corinne was screaming so loud, I wanted to cover her mouth to stop her.

Bern dragged Tommaso toward us by his armpits. His skin had swollen visibly, as if the bees had penetrated and were shoving from inside to free themselves. He had a double nose, a dozen eyelids, deformed lips, and an unrecognizable nipple among the red welts. When Giuliana, who had stayed where she was, saw him in front of her, her expression reflected back to each of us the horror that we hadn't fully realized.

I was the one to drive us to the hospital in Ostuni, ignoring traffic lights or right-of-way. Next to me Corinne stared straight ahead, more and more wide-eyed. She was no longer crying, but she didn't say a word. Bern and Danco had put Tommaso between them in the backseat, and Giuliana, before watching us speed away, had been quick to give them the knife we'd used to cut the bread. "Garlic! Bring us some garlic!" Bern had ordered, and after running around in circles, she had managed to get that, too. Now Bern was scraping Tommaso's skin with the blunt part of the blade, to extract the stingers. Danco, after peeling a clove of garlic, said, "Are you sure? It seems like a stupid peasant custom."

"Just rub it in!"

How many stings in all? Twenty? Thirty? "Fifty-eight," they told us at the hospital. The bees had even stung him on his scalp and inside an ear. There were bees trapped in his briefs; when they undressed him on the stretcher they rose up in flight. But we'd hear all this later from Bern, because he was the only one who followed the stretcher through the swinging doors to the emergency room. He was still wearing the paper coverall.

In the meantime, we were busy lying about the accident. No, we weren't raising bees, that required a permit, of course we knew it . . . Tommaso had stumbled upon a nest while cleaning the gutters . . . A very big nest, yes indeed, we had never seen one like that either . . .

Several hours passed before they told us that he was out of danger but sedated, and that they would keep him under observation. We spent all day and most of the night in the waiting room, sitting on plastic chairs bolted to the floor, under the fluorescent lights.

When it was all over and we were together again beneath the pergola, Danco lit into Tommaso: "Do you mind telling us what the fuck you were trying to do?"

"They came out suddenly."

"The fuck they did! Don't try to take us for a ride, Tommi. Did you put your hands inside the hives? What did you intend to do, huh?"

"I didn't put my hands inside the hives."

"Your shirt was unbuttoned!"

"That's enough now, Danco. Leave him alone," Bern intervened. For once, Danco obeyed him.

THAT YEAR the traps failed. Maybe the flies had spread the word around. After a furious argument we voted to buy the dimethoate, but it was too late for that too. The yield was miserable, the quality of the oil inferior. We didn't sell more than thirty bottles.

One morning we woke up without electricity. When Danco went to check the unit, he found the plates smeared with a mixture of glue and dirt. For hours we tried to figure out who could have decided to sabotage us. We had surrounded ourselves with enemies by taking field labor away from others.

The old power generator would no longer start, and we didn't make much of an effort to get it working again. For the first time we were seized by overwhelming dejection.

Corinne had an emotional meltdown. It took Tommaso almost an hour to quiet her down, as she kept on saying, "Are you taking responsibility for this? Making me stay in this cold with wet hair, now of all times?"

That evening Bern took me to our room and told me: "We have to ask Cosimo for help. Go talk to him. Ask him if we can tap into his power grid until we resolve this problem. We'll pay him for the extra consumption."

"He'll never agree. Remember how we treated him?"

"He won't be able to refuse a favor to you. He was so attached to your grandmother."

I decided to go alone. A fire must have been lit somewhere, the air smelled of woodsmoke.

I knocked at the door of the lodge and Rosa came to open it. She held the flaps of her robe closed and peered behind me, then let me in. Cosimo was watching television. When he saw me he tried to straighten his thinning hair that had been mussed up by leaning back in the chair.

I explained the incident regarding the solar panels, without admitting that someone had damaged them on purpose. Would he allow us to use his electricity for a while? Only for as long as it took to find a solution.

"Everything here is yours," he said gravely. "But you'll need an extension cord hundreds of feet long."

"The cables for the panels should be enough. If not, we'll connect them to other ones."

He looked up at me with a kindliness that I did not expect.

"You've become a capable young woman," he said. "There should be several feet of cable in the cellar."

"Thank you. We'll pay you."

I was ready to leave, but Cosimo took my hand.

"It's time to decide what to do with the villa, Teresa. Rosa and I continue keeping it in order, but it will deteriorate if no one lives there. And we can't go on doing it for free."

"I understand," I said, but only because I wanted to get back to the others.

Meanwhile, Rosa had prepared a basket with jars of preserves. "They're made my own way," she said. "I hope you and the others like them too."

Cosimo walked me to the gate.

"Those people," he said when we reached the boundary, "especially the curly-headed one . . ."

"Danco."

"It's none of my business. But you're a young lady who was brought up properly, Teresa. Those people are different. They grew up with roots that are too short. Sooner or later a gust of wind will pull them up and sweep them away."

But Cosimo didn't know what we knew: that plants grown in the safety of pots, with long roots that spiral tightly around, don't adapt to the soil. Only those whose roots are free, uprooted while young in winter, will make it. Like us.

"Tomorrow morning we'll come with the cable," I said. "You mustn't worry about anything." He nodded. In the dim light, he looked older.

A FEW DAYS LATER I confessed to the others that I was the owner of the villa. They were not angry as I'd expected, but strangely incredulous. They were silent for a few moments, then Danco asked: "How much is Cosimo offering?"

"One hundred and fifty thousand euros."

"That house is worth a lot more."

"I think that's all he has."

"That's his problem."

"What do you mean?"

But Giuliana ignored my question: "How much more, Danco?"

"At least double that, if I had to guess."

"So now you're also a real estate expert?" Corinne needled him.

He paid no attention to her. "It's falling down but it's vintage. And it comes with, what, seven or eight acres?"

I shook my head. I had no idea.

"He gave us the electricity," I said. By now I saw what he was driving at.

"We paid him for the electricity."

"But I promised him."

"And does this seem like the kind of situation in which promises are worth anything?"

I looked at Bern for support, but he said, "If your grandmother had wanted to leave him the house, she would have."

"What about all our talk about abolishing property?"

Danco gave me a sympathetic smile. "Perhaps you misunderstood certain aspects, Teresa. There is a radical difference between living equitably and acting like fools. We're not simpletons to be taken advantage of."

An excitement was spreading through the group, I sensed it.

"Teresa was keeping her treasure to herself," Giuliana muttered under her breath.

WE GOT in touch with an agency in Ostuni and a few weeks later I met the buyer, an architect from Milan, in the notary's office. Handing me the contract to be countersigned, he said: "Your grandmother's villa is magnificent, it must be costing you a lot to let it go. I promise you that I will fix it up so that it's good as new again, respecting its spirit."

"Thank you," I murmured.

Bern had accompanied me to the appointment, but he hadn't wanted to come in.

"This land is touched by grace," the architect said. Then, looking up from the contract, he added: "What can you tell me about the two custodians? Are they reliable people? I was thinking of keeping them."

A few days later Cosimo and Rosa left. And a week later the police came to the masseria. I wasn't really surprised when the officer, a young woman not much older than us, with a ponytail sticking out of her cap, told us that a report had been made regarding our unlawful presence there. What did we expect?

Tommaso and I watched her take a notebook out of her inside jacket pocket and flip through the pages.

"I understand there are six of you, is that right? I'd be grateful if you could call the others out."

When we were all gathered under the pergola, she asked us to provide our ID papers.

"And if we refuse?" Giuliana challenged her.

"You would have to follow us to the police station for further checking."

So we all went up to our rooms to dig up documents that attested to our being members of society after all.

"Will they arrest us?" I asked Bern in the few moments we were alone.

He kissed me on the temple. "Don't be silly."

The policewoman wrote down each one's details. In the meantime, her partner wandered around.

Giuliana tailed him, looking for any excuse to keep him away from the corner where the Super Skunk was planted. Just to distract him, she offered him a raw turnip that she tore from the ground; she eventually ate it herself, maybe to show that there was really nothing ridiculous about her overture.

Even worse than the waiting was my realization that this place, a place that for us was miraculous, aroused absolutely no sense of awe in the two strangers.

When the officer asked if anyone could claim a right to occupy the property, Bern spoke up. "The owner gave us permission to stay," he said.

She flipped through the notebook again. "You mean Mr. Belpanno?"

"He's my uncle."

Since I'd been there, that was the first time I'd heard him reaffirm his blood tie to Cesare.

"I spoke with Mr. Belpanno just this morning by phone. He wasn't aware that anyone was living here. The property is for sale and is supposed to be vacant, he said. Were you the ones who replaced the sign?"

"There was never any sign," Danco lied.

The policewoman noted his statement in her notebook. For the report she would write up against us, I thought. In a flash I could feel my parents' disapproval descend on me all the way from Turin.

"Do you have a warrant at least?" Giuliana asked sternly.

"We're not conducting a search, miss," the policewoman replied calmly. "In any case, if we did have a warrant we would be under no obligation to present it to you, given the look of things."

"There's been a misunderstanding," Bern interrupted in a firm voice. "Let me talk to my uncle and I'll prove it to you."

"Mr. Belpanno asked that the property be vacated within a week. Otherwise he will press charges."

She put the notebook on the table. Her tone when she continued was softer, as if she would like to be on our side, if she could: "Look, we have photographs. There is evidence of an illegal diversion of electricity from the power lines, a similarly unlawful installation of solar panels— which I would probably find myself if I were to walk in that direction," she said, pointing to the correct location, "and undeclared beehives, not to mention the marijuana plantation."

"'Plantation' is actually an exaggeration," Tommaso carelessly corrected her. We all turned to look at him. She pretended not to register that admission of guilt.

"My advice is that when we come back in a week, we find nobody here."

Corinne had slipped away into the house. She came back with two jars of honey and set them on the table in front of the police officers.

"Since you already know about it. It's a millefiori honey we produce ourselves."

"Now you're trying to bribe them with honey?" Danco asked angrily. "You're really an idiot."

The policewoman said: "I'm sure it's excellent, but we can't accept it."

A FEW MINUTES LATER we were alone again, the six of us, beneath our pergola, near the walls of our house, surrounded by our land, by everything that was ours and that suddenly no longer belonged to us.

Bern placed six beers on the table, but nobody reached out to take one.

"Stop acting like that, all of you."

Danco jumped on him. "It looks like you don't give a shit."

"We have the money that Teresa got from the sale of the villa. We can buy Cesare's masseria. It's for sale, isn't it? No more sneaking around."

"And how much would this Cesare want?" Danco asked skeptically.

"He'll accept whatever we offer him. Especially if it's us."

"It doesn't seem to me that your dear little uncle cares all that much about you."

Bern proposed putting the purchase of the masseria to a vote: "All in favor of making this property truly ours—forever—raise your hand."

I raised my hand, but I was the only one besides Bern.

"Well?" he persisted. "What's going on?"

At that point Corinne decided to grab a beer, uncapped it nervously with the bottom of her lighter, took a sip, then squeezed it between her hands.

"There's something we have to tell you," she started. "We thought we'd do it at another time, but given the circumstances, we might as well. Tommi and I are leaving. I'm pregnant."

She raised the bottle, as if to propose a sad toast. Tommaso was ashen.

"What do you mean, pregnant?" Bern asked, dumbfounded.

"Do you need me to explain it to you?"

But Bern didn't catch the sarcasm, because he was overcome by a surge of tenderness.

"Pregnant! That's great news! Don't you realize? It's the start of a new era. We'll have children. Teresa, Danco, Giuliana . . . don't you see? We'll have to get busy too. They'll all grow up together, here!"

The idyll that he had instantly imagined left his whole body quivering. He went behind Tommaso and Corinne and hugged them, then kissed each of them on the cheek.

"Pregnant!" he said again, not noticing that Tommaso seemed to be on the verge of tears.

"How many months?" Danco asked.

"Five," Corinne answered, looking at each of us in turn.

Bern wouldn't stop. "So what were you waiting for to tell us, huh? We don't need to vote now. We'll buy the masseria, we'll make it the perfect place for the kids. They'll have plenty of aunts and uncles and brothers and sisters."

That's when Corinne shook him off.

"Didn't you hear me? I said we're leaving, Bern. Going. Do you think I can let my child grow up in this place? For what, so he can get tuberculosis?"

It took Bern a few seconds to absorb the information.

"You're leaving," he said.

Corinne began toying with an earring. "My parents found us an apartment in Taranto. That way we're close and they can help us out. It's not very big, but it's right in the center."

"And what about us?" Bern asked.

Corinne lost her patience. "Holy Christ, Bern! You're really missing a few marbles, you know?"

But he wasn't looking at her anymore, he was staring at his brother instead, waiting for him to meet his eyes. Yet when he spoke his name in a whisper, then slightly louder, Tommaso didn't move.

So he came back and sat down next to me. He finished his beer in silence, then turned to Danco. "It's up to us four, apparently."

Danco puffed out his cheeks. "It would be mad to buy this place. Don't you see how run-down it is? The soil is poor. We have to work like hell here."

"What are you saying? It looks like you've all lost your minds today. We have the food forest here. We have the hens, the bees, everything."

Danco shook his head, as if he were inwardly fighting something.

"The police, Bern. I don't want to mess with them. And then, did you see what happened to the solar panels? And how things turned out with that shitty bastard Cosimo? We're not welcome here."

"None of us ever thought we were," Bern said.

I took his hand. It was cold and his fingers were trembling a little. I squeezed them. Danco rubbed his palms on his jeans. "What do you say, Giuli?" he asked. "I think it's time to clear out."

She snapped her fingers in reply, clearly conveying the idea that she hadn't expected anything else. Bern sat motionless, faced with that mutiny.

But Danco had something more to add: "I don't think it's fair to divide the money from the villa equally. After all, it was Teresa's. But each of us should get something, right? Kind of like a golden handshake. We all worked here, we all invested in it. What do you think, Teresa? You were the one who suggested putting the money in the communal kitty. Of course, now that things have changed, you can take back what you said, but . . . well . . . we all contributed."

No matter how hard he tried, he couldn't maintain his usual clarity, the objectivity that his scientific studies had taught him.

"I propose that those who leave receive twenty thousand euros and not lay claim to anything else. Twenty thousand each," he hastened to specify. "The rest will be left to Bern and Teresa. About one hundred thousand. That should be enough to purchase the masseria."

"And you just thought of this now?" Bern asked in a hard tone that he'd never used with Danco before.

"Does it make a difference?"

"Did you think of it just now or did you already do the calculation, Danco?"

Danco sighed. "Bern, people aren't property either."

"Don't you dare lecture me on morals."

Danco grumbled. "Whatever you say. So, Teresa, do you agree or not?"

"Teresa agrees," Bern answered for me. I squeezed his hand again.

"Good. So, what do you say, shall we make a toast to increasing the world's population? With decent wine, though."

Bern contained himself for the remaining time. He clinked glasses with everyone, even Danco. We pretended we were celebrating a new beginning, the birth and who knows what else, yet deep inside, each of us knew that the toast was mainly a confirmation of the end: the end of nights spent together under the pergola, the end of friendship itself maybe; the end of a cloudy dream that none of us, with the sole exception of Bern, had ever seriously believed could last.

———

THOSE FINAL DAYS. A virulent restlessness stirred in Bern. He spent a lot of time with Tommaso, the anguish of the new separation identical to that of the evening many years earlier at the Scalo. But this time their behavior was different. What they did was take walks together. Only once did I catch them hugging each other among the giant buds of cabbage scattered in the food forest: I didn't feel jealous as I had in the past, though, only deeply sorry for both of them.

The first to leave were Danco and Giuliana. They were headed south, they didn't know where exactly. In front of the jam-packed jeep, Danco suggested one last time that Bern follow him. I held my breath before he answered, worried that the painfulness of that moment would lead him to agree. Instead he shook Danco's hand and said, "If I move away from this place, I'll die. I know that now."

With two days left before the deadline issued by the police, Bern and I remained alone. We sat on the bench under the holly oak. It hadn't been used in quite a while, because it could accommodate only two people at a time. Bern held me close to him. The countryside was so silent and still that we felt like the last human beings on earth, or the first ones. Bern must have had a similar thought, because he said: "Adam and Eve."

"The apple tree is missing."

"Cesare claimed that it was actually a pomegranate tree."

"Then we have one."

His chest rose and fell. Then his fingers slid over mine, gently, looking for a way under the sleeve.

"We'll go to him tomorrow," he said. "We'll make him an offer for the masseria."

"There won't be any money left, afterward."

"Who cares?"

I looked at the land around us. The realization of the work that from then on would fall on just the two of us was daunting. If in some recess of my mind I still imagined resuming my studies, connecting my earlier life to the present one as though grafting two cuttings, that was the moment I realized that it would never happen. It was Bern and me and the masseria and nothing else. I was twenty-five years old and I didn't know if that was too old or too young to live like this, nor did I care to know. I loved Bern more than ever at that moment, as if our sudden solitude had allowed that feeling to finally expand and occupy everything.

So when he said, "We must have a child, like Tommaso and Corinne"—not "I'd like" or "We could," but "We must," as if there were no other way—when he said that I was sure he was right, and I replied: "We will."

"Tonight?"

"Right now."

But several more minutes passed before we made up our minds to move, to walk into the house and go upstairs. And in that silent interlude under the holly oak, we envisioned the image of a little girl, our little girl—who knows why a girl—dancing a few steps away from us, picking a dandelion from the ephemeral grass and offering it to us. It was nothing but a fantasy and we didn't admit it to each other, not even later, but I was certain then, as I am certain today, that we saw her vividly before us, and that our visions were identical. Because at that time that kind of thing happened between Bern and me: we used fewer and fewer words, but we were still able to recognize the visible and, as if by tacit accord, even conceive of the invisible.

4.

I found Bern painting a picture on one of the outer walls of the house, the one facing north. Dark brushstrokes in a glossy brown left over from restoring the doors; the paint stood out on the rough whitewashed surface. Mornings were already cold, with lots of dew. I pulled the collar of my sweater up over my chin.

"Yep, it's a penis," he confirmed without turning around.

"So it seems," I said, trying not to look surprised. "A huge penis on the wall of the house. The neighbors will love it."

"In Tibet it's considered auspicious."

Only then did I notice the illustrated book resting on the ground, no doubt borrowed from the Ostuni library, where Bern sometimes disappeared for entire afternoons. He was copying the image from there.

I went closer, to compare the photograph with the result. Bern's version was too simplistic, it looked more like a kid's vulgar vandalism than it did the original.

"So we're back to magical thinking?" I asked, placing a hand on his shoulder.

He gave a lopsided smile. "I told myself it doesn't hurt to try. We'll attract some benevolent spirit. For our cause."

Our cause: the phantom daughter who by then had taken over every conversation, every thought and desire. It had been almost two years since the afternoon when we'd first imagined her, when we'd pursued her like a hallucination, up the stairs and into our bed to make her real.

There was already a room prepared for her upstairs, the room that had once been Tommaso and Corinne's, and before that, Cesare and Floriana's. Bern had carved a cradle from the stump of an olive tree, but the cradle stood empty, in the center of a room just as bare.

"You could help me," he said. "You're better at drawing than I am."

I took the can of paint and the brush and tried to correct the outlines. Bern looked on over my shoulder.

"That's much better," he said finally.

"Who knows what people will think."

"What they think doesn't matter. And besides, who? Nobody ever comes here."

It was true. Not even Tommaso and Corinne came anymore. Since they'd had Ada, they lived barricaded in the attic financed by her father, exhausted by getting up at night but content as could be. We went to visit them often enough, but less and less willingly since our failure had become a chronic disorder. Even when we decided not to drive up to Taranto and endure the sting of envy, Ada's accomplishments were reported to us by phone. Ada who stood up, holding on to the bars of her crib. Ada who waved her hand to say ciao. Ada who was getting her baby teeth.

Danco and Giuliana appeared less and less frequently as well. So there we were, Bern and I, property owners now, still young but extremely disheartened, worshipping a pagan totem.

I said: "Maybe it will work."

"Let's hope."

"Or maybe it might be time to see a doctor, Bern."

He whirled around to face me. "A doctor, what for?"

"Maybe there's a problem. With me."

"There is no problem. We just have to keep trying."

He took me by the hand and we went into the house. I made us some breakfast. In November there were the starlings; they made raids on the olive trees. We heard the shot of a hunter in the distance. From the window, I saw the black formation of birds fan out for a few seconds, startled, then regroup as if nothing had happened.

THE PAINTING on the wall did not help. My periods kept coming with ruthless punctuality, and each time Bern was more disappointed and wrought up. I got to the point of hiding the tampons from him, but he found out anyway when he pressed his chest to my back at night to make another attempt. We can't, I'd tell him without turning around; then he'd slump back on the mattress and begin calculating how many days it would be before we could try again.

Sex was the thing that changed the most. Before, we were wild savages, whereas now Bern's thrusts had a martial regularity, as if he were searching for an exact spot inside me. Before, even after he came, his fingers didn't stop until my stomach started jerking uncontrollably, whereas now he withdrew immediately, as if he didn't want to disturb the biological process that was under way. Before, we'd lie beside each other, exhausted and drained, whereas now he made me keep my pelvis elevated for ten minutes. He timed them on the clock.

We did not know a specialist who could help us, so we went to the bar in Speziale to consult the telephone directory. We copied down the numbers of four or five gynecologists near Brindisi, glancing around as we did so, as if everyone there could tell what we were doing.

We went back to the masseria to make the call. Bern let me choose among the names. Walking in circles between the holly oak and the house, I explained our situation to the doctor, the months of failed attempts. Spoken aloud, the fears that until now had remained vague suddenly became concrete. The doctor asked me questions, questions that in the following weeks would become obvious, but that during that first conversation sounded like accusations: our ages (twenty-seven and twenty-eight), previous illnesses (none), the characteristics of my cycle (regular, heavy), the presence of abnormal bleeding (none), when we had stopped using contraception (about two years ago), and what strange reason had made us wait all that time to call.

In any case, the doctor said finally, he didn't treat fertility; he had me take down the number of one of his colleagues, a Dr. Sanfelice, not in Brindisi, in Francavilla Fontana; we could say he recommended us.

So I repeated the call from the beginning, trying to sound more confident though I was slowly losing my courage to go through with this. The same questions and the same answers, in nearly the same order, as I kept walking between the holly oak and the house, pivoting around Bern, who registered every word, silently encouraging me.

The next day we were in Dr. Sanfelice's waiting room, properly dressed, as if a good result depended on first impressions. On the wall hung a print of the female reproductive system, with black lines connecting the organs to their names. There were two other couples, only one with a big belly. Both women gave me empathetic smiles, maybe they could tell I was there for the first time.

Sanfelice had me lie down on the examining table, slipped on a latex glove, and told me to relax as he gave me a little slap on the buttock. As he moved the probe, he spoke incessantly. The only information about us that he'd noted, or maybe the only one that intrigued him, was that we lived in the country. He had a house outside the city himself, he said, then went on to talk about oil pressing.

When we were sitting in front of his desk again, he asked about the frequency of our relations. "You have no idea how many couples come here saying: Doctor, we've been trying for a year. And when I ask: How many times a year? They tell me: At least five or six!"

He laughed, as if at the punch line of a joke, but he composed himself almost immediately, maybe because we'd remained serious.

"I ask partly because, at first glance, everything seems to be in order with the lady."

"Every day," Bern said.

"Every day?" The doctor's eyes widened. "For more than a year?"

"Yes."

Sanfelice frowned. He toyed with a magnifying glass, then put it back in its place. He turned to me. "Then we should investigate more thoroughly."

"What could it be?" Bern asked.

"Your sperm could be slow or meager, or even slow and meager. It could be your wife's ovaries, even if there are no fibroids. Endometriosis, at worst. But it's not worth talking about until we've done a nice round of tests."

He began filling out the authorizations, going on for quite some time. Bern stared at the doctor's hands.

"Come back when you have everything," he said, handing me the forms. "I haven't written one for a spermiogram because that can be done here at our office. The collection is on Tuesday, here are the instructions," he said, handing us another preprinted sheet. "It costs one hundred and twenty euros. Check around if you like, but you won't find a better price elsewhere."

"Can this be resolved, Doctor?" Bern asked when we were already standing.

"Of course it can be resolved. We're in the twenty-first century. There's almost nothing that medicine can't do anymore!"

Along the brightly lit streets of Francavilla's center, people were strolling in and out of shops, ducking into bars for the aperitivo hour. A stall was selling candied orange peels; I asked Bern to buy a bag, but he said: "Let's go to a restaurant!"

We had never been to one by ourselves, he and I. A strange anxiety came over me, as if I weren't ready.

"We have all those tests to pay for."

"But didn't you hear what Sanfelice said? There's nothing that can't be done. Soon we'll have our baby girl. We have to celebrate! I should have listened to you and come here sooner. You pick a place."

I spun around in the middle of the piazza, as amazed as a young girl seeing the city, the vivid streetlights, and the baroque buildings for the first time.

"That one." I pointed.

I clung to his arm, ecstatic, as though we were on the first date we'd never had. I let him sweep me into the restaurant. We were just another couple in love, like any others, at least for that evening.

THE RESULT of all the expensive tests was: absolutely nothing. There was nothing wrong with his sperm count. And there was nothing wrong, at least so it appeared, with my levels of progesterone and prolactin and estradiol, with the LH, the TSH, the FSH, all those acronyms whose meanings were still unknown to me. Yet I did not get pregnant. As if something—which was what Dr. Sanfelice thought, though he didn't dare suggest it—wasn't working with Bern and me together.

"Let's get this over with," the doctor said, contemplating the conspicuous display of reports on his desk. "A round of insemination and the problem is solved."

First, however, it was necessary to go through ovarian stimulation. A strict schedule of times and administrations: for that too, Sanfelice had a preprinted sheet ready, which he handed me with an encouraging smile.

During that same period, Bern decided to rebuild the treehouse in the mulberry, exactly where it had been before. Our little girl would like it, he said. He talked about the project as if it were a top priority. It was useless to try to make him see reason, to remind him that in the best of cases our daughter wouldn't climb that tree until she was four or five years old. He would turn up at the masseria with a load of wood planks, then disappear into the fields for hours, looking for flexible branches with which to construct the roof.

In reality, he couldn't stand being idle while I was urging my ovaries to produce more, and still more, until I reached the point of collapse. And while he soared up there in his dream of fatherhood, I was crushed to the ground by my fattened stomach, my hard breasts, the patches of cellulite that appeared overnight on my thighs.

"Don't look at me," I'd say when it was time to get undressed in the evening.

"Why shouldn't I? I always look at you."

And yet I knew he couldn't prevent his gaze from analyzing me, from registering every one of those signs of deterioration.

"Just don't look at me, that's all."

His solicitude got on my nerves at least as much as it supported me. It made me feel even more trapped, even less desirable.

"I wish it were me undergoing the treatment," he said.

"But it isn't." Then, remorseful, I added: "Be satisfied with your vitamins."

Sanfelice had prescribed them to improve the quality of his semen. I doubted they really did anything, but Bern took them scrupulously, as if our whole mission depended on them.

ONE DAY Nicola showed up at the masseria. We had lost touch some time ago. The only news I had of him came from the brief encounter I'd had with Floriana over the sale of the property, a couple of years back. She'd been laconic: he's fine, she'd said, and I hadn't dared insist.

He arrived on a bright Sunday morning in May. He got out of a shiny, smart-looking sports car, and he too was smartly dressed: he wore leather shoes and an immaculate white shirt, slightly open over his tanned chest. He had filled out since the last time I'd seen him, bulked up for the better, I thought. He looked more like Cesare, the same muscular good looks, even a trace of his glow.

Mentally, I listed how slovenly I must look to him: my hair tied back and a little dirty, Bern's shorts that I wore for working in the garden, the film of sweat on my forehead, and the clearly visible stain under my armpits. My skin oozing gonadotropins.

"I hope I'm not disturbing you," he said. "I was passing through the area."

"I'm the only one here," I replied, assuming he was there to see Bern.

Nicola looked around, hands on his hips and with a pleased expression. "Cesare said I'd find everything somewhat changed. But it doesn't seem so different to me. Even the swing-chair is still there."

"Don't sit on it, it's unsafe. We made some changes inside. And the vegetable garden, that part is all new. How about some lemonade? I'll bring you some."

When I came back out, Nicola was sitting at the table, typing a message on his phone. He put it away and finished the lemonade in one gulp. I poured him some more.

He pointed to something on the side of the house with an amused expression. The fertility mural had been there for so long that I didn't even notice it anymore. We'd covered it with a coat of whitewash, but the dark outline had reemerged as soon as the paint had dried.

"A bet," I explained, certainly turning crimson.

"A lost bet, I guess," Nicola said.

I had never felt self-conscious in his presence. He had always been the awkward one. But in becoming adults there had been an imperceptible shift.

"Are you still with the police?" I asked.

"Special Agent Belpanno, at your service," he said, showing me a tiny, gold-plated badge on his shirt. If Danco had been there, he would have laughed in his face.

Nicola revolved the glass a half turn, a gesture that reminded me of how he'd been as a boy.

"Cesare couldn't accept it at the beginning. Because of the weapons, you know. But then he realized that weapons have little to do with it. It's about having a certain kind of ideal." He paused, as if reflecting on what he had just said. Then he shook his head. "I'm not cut out for the type of freedom that he preaches. And what about you? Do you like being here?"

I crossed my arms over my chest.

"It's hard. Maintaining the masseria and all the rest. Sometimes I feel like I've become part of the landscape. Like a plant or an animal."

Why was I confiding all this to him?

"But I can't imagine a different life," I added.

"You and Bern should come to the city once in a while. I even have a guest room. I'd like to introduce you to Stella."

"Is she your girlfriend?"

"For two years now. But we don't live together."

He waited for me to accept the invitation, or turn it down, something. Bern and I in Bari, visiting him at his house.

"Do you mind?" he asked.

"About what?"

"About Stella. That we're together."

I straightened my chair. "Why should I mind?"

"There wouldn't be anything wrong with it. It bothered me when I heard about you and Bern."

"I'm happy for you," I said. "Would you like some cookies? I'm experimenting with almond flour. They aren't great, but they're edible."

Nicola waited quietly for me to bring them outside. He took one from the plate, and when he took a bite it fell apart.

"They're too crumbly, I know."

He smiled. "You just need to know the technique."

All that time apart and we had already run out of things to say. No, that wasn't it. There was the past to talk about: when we played cards on that same table, the intricate tangle of attractions that bound us as kids, the time he'd given me a coral bracelet that I had never worn yet still had, why I stopped answering his letters. But it was too dangerous, we both sensed it.

"Bern and I want to have a child," I said.

It popped out all of a sudden, the words followed by a backwash of shame.

"I'm undergoing treatment. I'm taking hormones."

"I'm sorry," Nicola said in a soft voice.

Everything pent up in me suddenly came rushing out. Tears filled my eyes. "The test results are fine, but nothing happens."

I had embarrassed him. And saddened him. And irritated him, probably.

"A colleague of mine had a varicocele and for that—"

"Here comes Bern," I interrupted him.

Nicola turned in his chair. He raised a hand to him, but Bern did not respond to the greeting. We watched him walk up the dirt track. The tears kept coming; I couldn't stop them, and for some reason I didn't even want to. I just wiped them on my wrist.

"What are you doing here? Teresa, did you invite him? Why did you come?"

I stood up and took Bern's hand. "He stopped by to see us. We hadn't seen him for so long. I offered him some lemonade."

Nicola watched us with an unreadable expression.

Bern was extremely edgy. "Why are you crying? What did you talk about?"

His eyes went to Nicola. "What, huh?"

"Nothing," Nicola replied, holding his gaze.

Bern would not have forgiven me for telling him about the treatment.

"You have to leave," he threatened Nicola. "This isn't your place anymore. We paid for it, understand? Go!"

Nicola got up slowly. He positioned the chair under the table, then took another look around, as if to take in the wonder of the masseria one last time.

"It was great to see you," he said to me finally.

He took Bern by the shoulders in a kind of embrace and touched his cheek to his. He gently touched his beard, maybe because he'd never seen it so long. Bern remained motionless, offering no resistance. Then Nicola got in the car. He drove off, honking the horn twice as he reversed.

I picked up the lemonade pitcher, but I didn't know what to do with it and I put it back where it was.

"Why did you treat him like that?"

"He has no right to come here," said Bern, who had sat down, his eyes fixed on the empty table.

"You were like brothers. Now you and Tommaso act as if he'd never existed."

His thumbnail dug into the plastic tablecloth.

"Cops like him should stay away from this place."

"You chased him off as if he were a criminal. You were the one who acted like a cop!"

Bern bowed his head. "Don't be mad at me. Please."

His voice was so defenseless, so sweet. My anger was swept away in an instant, replaced ounce per ounce by the usual surge of devotion.

I sat down, put my arm on the table, and rested my head on it. Bern's fingers were quick to find their way into my hair.

"We're very tired," he said, "but soon things will be different."

His fingers massaged the roots, rhythmically. Eyes closed, the late May sun on my eyelids, the silence of the countryside: I let all those things envelop me, like a new promise.

IN MY SECOND YEAR of high school, the family doctor had cauterized a wart under my big toe. Before starting, he'd said, "Now we'll do some welding on this little girl." My father was squeezing my hand and telling me not to look down, to keep talking to him. It was the only clinical procedure I had ever been subjected to. So on the day of the oocyte retrieval, while I slowly took my clothes off behind the screen and put on the coarse, demeaning paper gown, hearing Bern's and Sanfelice's voices out there, my body was trembling, as if a sudden chill had entered the room.

The procedure was brief, however. The doctor commented step by step on his miraculous probing inside my anesthetized cavities. It was a way to reassure me, but I would have preferred him to be silent. I saw his assistant smile kindly behind her mask, a girl my age, who in all likelihood would never have to undergo such a treatment. For some time I'd been dividing women into two categories: those who could easily conceive and the others, like me.

"Nine!" exclaimed Sanfelice, handing her the probe.

"Nine what?" Bern asked him, mesmerized by the deftness with which Sanfelice slipped off his gloves, stretched his fingers, and scribbled something on the chart.

"Nine follicles. We'll have so many oocytes that we can make a nestful. Excellent job, Teresa."

Through the sheet he gave me the same little slap on the buttock as he had the first time. Before the pickup he had started calling me by my first name, because we were allies now, he and I, in the front line of that battle.

The next phase would take place in the laboratory, under the lens of a microscope. There, far removed from our eyes, in a silent coitus in the midst of a perfectly sterile environment, Bern's fluid would be mingled with mine. Nature would take care of the rest, even if I no longer used that word "nature," at least not in front of Sanfelice, after he'd gotten angry with me right in the middle of the dilation: "Natural?" he'd snapped. "What would be natural in your opinion, Teresa? Are the clothes you wear natural? Is the food you eat natural? Oh, of course, I know you grow your own vegetables, the ones you brought me last time were excellent. And you probably don't use pesticides or products like that, but if you think your tomatoes are natural, then, forgive me for being frank, you're naive. Nothing natural has existed on this earth for hundreds of years. Everything is the result of manipulation. Everything. And you know what I say? Praise be to God for this, even if he doesn't exist. Because otherwise we would still be dying from smallpox, from malaria, and from bubonic plague, not to mention childbirth."

Bern had not contested that outburst, not even afterward. I wondered if he remembered what Fukuoka had to say about medicine and doctors. But no, not even Fukuoka existed anymore, swept away by longing, and by an unconditional trust in Sanfelice and his techniques.

Outside the clinic, after the retrieval, I nearly collapsed. I hadn't eaten anything since the night before, not even the sugary tea the doctor had recommended. Bern held me up before I could fall.

"It's all those drugs," I whimpered.

He kissed me right there on the sidewalk, with people passing by who knew nothing about us.

"It's over," he promised.

That evening, in fact, I felt lighter, the local anesthesia was wearing off, my legs slowly began to feel like my own again, and the exhaustion of the previous days diminished, though I had not stopped the hormones. It was the thought of our baby that raised my spirits. Maybe she already existed under the microscope and would soon be inside me.

Sanfelice's assistant called the next day to summon us to the office. She wouldn't say why. On the way to Francavilla we were so distressed by what that phone call might portend that we did not exchange a single word.

Sanfelice was in a good mood, spirited, even, as he told us that the nine follicles, so promising, that he had congratulated us on less than twenty-four hours earlier were empty, not even one oocyte among them.

As always, my understanding struggled along behind his words. I asked, "How is that possible?" as I felt the emptiness Sanfelice had spoken of spread through my abdomen, through my chest and throat.

"Anything is possible."

He had a nervous tic that made him blink his eyes and then reopen them in a kind of amazement. He did this twice in a row, before adding: "We are operating in the realm of statistics, Teresa. But I'm planning to change the treatment. In place of the Decapeptyl, which you told me you did not tolerate well, we'll try the Gonal-F used in combination with Luveris. Had I already given you the Luveris? No, in fact. And we'll increase the dosages a little."

"Another stimulation?" I asked, already tearful. I wept with shameful ease in those weeks. The explanatory leaflet, which I had read and reread, said that could happen.

"Come, now, my dear!" the doctor urged me. There was a hint of irritation in his voice. "Sometimes it's necessary to accept a little sacrifice for a good result, am I right?"

He repeated it. "Am I right?"

Bern nodded for me.

Then we were out on the street again, on that corner of Francavilla that would become the backdrop of all the memories related to those months. There was a fruit-and-vegetable store opposite the entrance to the clinic. The vendor always stood outside, leaning against the doorpost, watching those who went in and out. Who knows if he was aware of what was going on.

"I don't know if I can do it," I said to Bern.

"Of course you can do it."

He was already leading me toward the pharmacy for the new medicines we needed, new ways to force nature, whatever it was, to do what it didn't seem to have any intention of doing.

THE SECOND STIMULATION CYCLE was pure hell. My abdomen, my hips, my back, my legs—every single muscle hurt. I hardly crawled out of bed anymore. I stayed confined to our room, which had been turned into a field infirmary: old medicines and new ones were piled up everywhere, along with open packets of disposable syringes and glasses bearing the residue of a soluble powder that Sanfelice had prescribed by phone for headaches.

Bern was incapable of tackling that mess. During the day he managed to keep up with all the work in the fields by himself; I was afraid he'd get one of his back spasms, then we'd really be in trouble. Between one chore and another he poked his head into the room to ask me if I was any better. He never asked how I was, only if I was better, then he disappeared, afraid of the answer. In the evening, exhausted, he fell asleep on the edge of the bed to leave me as much room as possible.

One night the cramps were so strong that I woke him up. He didn't know what to do. He went downstairs and came back with a pot of boiling water, as if I were about to give birth. I screamed at him, so he vanished again and reappeared with a basin of cold water. He dipped the corner of his T-shirt in it and rubbed my forehead.

"Don't grind your teeth like that," he pleaded.

I told him that maybe I was dying, and he started shaking his head, stricken with panic.

"You can't," he repeated, "you can't."

He wanted to call an ambulance, but he would have had to walk to the end of the dirt track and farther on to the intersection with the paved road, leaving me alone all that time. The ambulance would not have been able to find us otherwise.

He pounded his thigh with his fist, as if trying to transfer the pain to himself. I told him to stop it. A great calmness had suddenly come over me, along with a kind of pity, not for me, solely for him, for his fear-ravaged face.

Finally I fell asleep. When I opened my eyes again the room was flooded with sunlight. Bern was still there beside me. He had picked some chervil flowers and put them in a jar with a sprig of bay laurel, which he'd set on the bedside table. He stroked my head and I slid closer to him.

"I spoke to Sanfelice," he said. "You have to discontinue the treatment immediately."

He couldn't look at me.

"There are still six days left."

"You have to discontinue it," he repeated.

"I made too much of it last night. I'm sorry. But I'll do better now, I'm sure of it."

Bern shook his head. His entire universe had come crashing down. I looked at him, his eyelids reddened by lack of sleep, his beard so long it curled, and the sense of defeat that seemed to weigh so heavily on him.

All that remained of the night's ordeal was a strange lucidity. Maybe I had dreamed something, though I could only dimly remember it. "You're not the problem," I said.

He didn't turn to look at me. But his shoulders stiffened for a moment.

"You don't have to—"

"There's another solution," he interrupted me. "Sanfelice wants to explain it to us face-to-face. Get dressed, we're going to see him."

"I HAVE COLLABORATED with this clinic for many years," the doctor told us. "It's in Kiev. Have you ever been there? A delightful city, where everything is very inexpensive."

He waited for us to shake our heads.

Kiev.

"I and my colleague there, Dr. Fedecko, who is very highly regarded in the fertility field, take on situations, how to put it . . . situations that traditional assisted fertilization cannot resolve. And at this point I would say that is the case here, despite your young age. It may be that the lady is afflicted with empty follicle syndrome. It is somewhat unusual, to be truthful, but not extremely rare. And in any case, we cannot ascertain it because she does not seem to tolerate the ovarian stimulation. Am I correct?"

He looked at me intently, as if he were expecting a disavowal just then, an admission that the agony of the night before had been an exaggeration, an act.

"Right," he continued. "We cannot afford the risk of hyperstimulation. So the only remaining solution is egg donation."

"Meaning the child will not be mine," I said softly.

Bern didn't understand. He looked at me, then at Sanfelice, then back to me. He had not read all the things that I had read in the past few weeks. The illusion persisted in him that this process was nothing but a way to hasten what would happen in any case. Innocuous, like his vitamins.

"Such nonsense, Teresa," Sanfelice said, joining his hands. "That's what everyone thinks at first. Do you have any idea how many of the children you see around have been conceived that way? Go ask their mothers if they aren't their children."

He leaned toward me.

"Children belong to those who carry them in their wombs. To those who give birth to them and then raise them. Do you know what the most recent studies say? We're talking about American studies, published in *The Lancet*. They say that the fetus assumes an unimaginable number of the pregnant woman's traits, even if it does not share her genetic patrimony. Unimaginable."

"Why wouldn't it share the genetic patrimony?" asked Bern, who was increasingly bewildered.

But neither Sanfelice nor I answered him. I was still caught up in that word, "pregnant."

"Do you know what happens in fact? Women come back here years later just to tell me: Doctor, my son looks like me. He looks more like me than like his father. And I say: But why does that surprise you? Didn't I promise you he would? For the donors we respect all the primary parameters: height, eye and hair color. The young woman will likely be a kind of dead ringer, even if you will never see each other. If instead you prefer a child with red hair or one who will be very tall, that's fine too, we'll look for a suitable donor. One of my patients insisted on having a mulatto baby and we satisfied her. If you could see the little girl, with that caffelatte complexion! She's already going to school."

Selected as though from a catalog, I thought.

Sanfelice spoke to Bern again. "Then too, the Ukrainian women are a treat for the eyes. Everyone thinks immediately 'Russian,' but it's not so. They don't have those Slavic traits, they're much more similar to us."

He leaned back in his chair, waiting for our questions. But we were both too shaken to speak, so it was he who once again broke the silence: "There won't be a religious problem, I hope. Because I have excellent reasons to present to you in that case as well. For one thing, Orthodox Jews from Israel come to Fedecko's clinic. And many Muslims. You can't imagine the fertility problems they have over there."

"Is it illegal?" I asked.

Sanfelice pouted.

"What can I say? It takes time to change people's ideas, especially here. If you're asking me what could happen once you have a beautiful, healthy embryo implanted in your uterus, whether someone could come and claim it, then the answer is no. Everything that grows in your belly is yours. And by that time the trip to Kiev will have been forgotten. Except when you feel like returning to me to have another one."

He turned from side to side in his swivel chair, his arms outspread.

"Can you imagine when all this didn't exist? We live in an age of infinite possibilities!"

After that he began a detailed explanation of the procedure and timing, and the new hormonal regimen, much milder than the previous one, "a piece of cake." The great advantage was that now I only had to prepare myself to be an "envelope."

An envelope.

I again lost the thread of what he was saying. What did I know about Ukraine? Only the Chernobyl disaster, the stories from my mother, who had stopped buying fresh milk because people said the cows were radioactive. I imagined abandoned gray villages, annihilated grain fields under a leaden sky.

Bern had squirmed forward in his chair to lean toward Sanfelice, drawn by the magnet of his knowledge. He soaked up the man's words as if they were magic formulas.

"We make every effort to keep costs to a minimum," the doctor finally said. "Eight thousand euros is all-inclusive. Plus the cost of the flights and the hotel, of course."

But eight thousand euros was much more than we had set aside. Our last savings had gone in the failed attempt at insemination. Now we barely had a thousand.

Bern and I looked at each other for the first time since the start of the visit. And from that moment on our anxiety was focused elsewhere, once again on how to get the money, almost as if the decision itself, whether to have the donor egg implanted or not, whether it was right or immoral and unprincipled, wasn't really relevant. It wasn't even worth thinking about. After all, what other choice did we have?

Eight thousand euros. Adding the expense of airline tickets, hotel, and meals in Kiev, the cost would be close to ten thousand. There was no way to quickly scrape together such a sum. With the very thin margins of what we sold at the market it would take us two, maybe three years; and there were the contingencies to consider, everything that was constantly breaking down at the masseria, not to mention the crops that could be wiped out overnight by hail, frost, or moles.

The doctor said that we were now in the twenty-first century, the age of infinite possibilities, when men and women with lab coats and sterile gloves in silent rooms, in Kiev, could take care of what we were not able to do. But Bern and I were still living in an era much further back; we were at the mercy of the sun and the rain and the seasons.

We knew of a money lender in Pezze di Greco, but he had the reputation of being a shark. We dropped that idea.

Without telling Bern, I phoned my father. I was always the one to call, though I did so rarely. We were speaking to each other again, but he still acted as if I were living in an inaccessible corner of the world. There was a flicker of surprise in his voice, though he soon reverted to his usual terseness.

"If you could lend me some money," I said without beating around the bush, "I'll be able to pay it all back after the olive harvest."

"How much do you need?"

"Ten thousand euros. We have to fix the roof."

I was surprised at how little it cost me to lie to him. He sighed into the phone.

"There's your college tuition to pay," he said. "The bill was sent to the house."

"You don't have to pay my college tuition."

I felt slightly breathless. My body hadn't completely recovered from the hormonal treatment.

"We need the money for the roof now, Dad."

"Your grandmother's house had a nice solid roof."

"I'm sorry, I already told you."

"Don't expect a cent from me. And don't you dare ask your mother. I'd hear about it, in any case."

Then he hung up. I stood still for a few seconds, my hand holding the phone uselessly to my ear. I had the bizarre sensation that the land around the masseria had suddenly expanded hundreds of miles in every direction, leaving Bern and me alone in the middle of a deserted plain.

It was my disappointment that made me confess what I had tried to do. We were in bed. Bern did not get angry as I'd feared; he didn't say the unkind words about my father that I'd expected. Instead he was silent, his eyes narrowed to focus on an idea that became gradually more defined. Then he smiled tightly. In a way, I was the one who gave him the idea.

"Your parents are respectable people," he said. "They follow conventions. All we have to do is come up with a situation that they can't turn away from, out of decency."

"Is there one?"

"Of course there is. Don't you see?"

"No."

"Marry me, Teresa."

And even in that absurd context, even in the preoccupied way Bern pronounced those vital words, as if he wasn't considering their meaning at all but something much more important that would come from it, the effect on me was an intense tingling in my cheeks that quickly spread throughout my body.

"We've always said that marriage is a social bond, Bern. We talked about it with Danco."

I was afraid the conventional girl in me would spring out from behind that show of composure, as if Danco were physically present in the room, analyzing the traces of foolish emotion that transformed my expression.

Bern slid out from the sheets and knelt on the bed, half naked as he was, his hair disheveled.

"If we get married, they'll be forced to give us a present."

"You want to turn our wedding into a fundraiser?"

"It will be a celebration, Teresa! Here. All the trees bedecked with white ribbons. And afterward we can go to Kiev. Up you go! Get up, quick!"

I pushed the sheet aside and stood up on the mattress. Bern was on his knees in front of me. Looking down on him, his close-set eyes had an even greater effect on me; I thought they'd been purposely created that way so he could pronounce the words that followed: "Teresa Gasparro, will you be my wife?"

I grabbed his head and pulled him to me, his ear pressed against my navel, so that he could hear the answer that had been lodged there for years, could hear it well up from the cavern where it had been waiting all that time: More than anything in the world, yes.

THE FACT THAT it was just an expedient, a performance, a scam, didn't matter. I believed in that wedding with every ounce of faith in me. The promise that we'd exchanged was enough to eclipse even the thought of the terrifying trip to Kiev waiting in the background. I didn't dwell on it, I did everything I could to put it out of my mind. For the first time in months, I was happy again.

The first list of invitees did not exceed fifty people. Not enough, even supposing an extraordinary generosity on their part. We began to extend it, first including relationships that had weakened, then those virtually forgotten. It still wasn't enough. The list was expanded, mainly from my side. I'd rack my brains for a name and suggest it to Bern: "The Varettos."

"Who are . . . ?"

"Friends of my parents'. They came to dinner every now and then. And one year I went to summer camp with the daughter. Ginevra. Or maybe Benedetta."

"Then let's add her, too. Does she have a boyfriend?"

"We'll put 'and guest.'"

I doubted it would work, but Bern seemed sure of it: "Everyone loves parties. Weddings especially."

And just as he had predicted, the support was astounding. Of the nearly two hundred people who'd been invited, almost a hundred and fifty replied that they would come, even though the wedding would take place so far away, and even though it was such short notice. So soon, in September? Yes, we had acted on impulse. Occasionally they did not conceal their surprise, we hadn't seen each other in so long . . . But I thought of you often over the years, and saying so, I almost convinced myself that it was true. They were moved. Will it be in a church? No, we decided on a civil ceremony, Bern and I are a little skeptical about religion.

Then I touched on the most delicate topic: we chose to forgo gifts, we don't really need anything, but we dream of taking a trip, far away, we haven't yet made up our minds exactly where. There will be an amphora in which to leave whatever you'd like.

We started extolling the appeal of the Apulian countryside, the light and the sea in the month of September. On that score at least we weren't lying.

Bern agreed to calling Cesare, Floriana, and Nicola as well, without protest, as if any resentment were suddenly forgotten.

"There's also our neighbor," Bern said when we had exhausted all our ideas, "the guy who bought the villa."

"The architect? I never saw him again after that."

"Then go pay him a visit."

So I picked some vegetables, and with a full basket walked over to the entrance to the villa.

The paved patio area had been enlarged and every tree was surrounded by a flower bed. The caretaker's lodge was unrecognizable, enclosed by a long window that was dazzling in the afternoon sun. No more damp patches, no more peeling. I wondered if my grandmother would have liked it. A wall about six feet high had been erected around the courtyard, a kind of fortification, which enclosed the pool and kept out the fields.

"It's because in the evening I get a little scared," the architect said, joining me out there. "I'm impressionable."

"I hope I'm not disturbing you."

"On the contrary. I was hoping that sooner or later you would come and see what I've done here. Teresa, isn't it? I'm Riccardo."

"I remember."

I handed him the basket of vegetables. The gift seemed foolish and inappropriate for someone who lived there, but Riccardo marveled at it. He set it down on a precise spot on the pavement, in the shade, and photographed it with his phone, searching awhile for the best angle.

"It's a perfect color composition," he said, "I'll use it for my blog."

"You can also eat them, afterward."

"You're right. Of course."

He offered me a tour of the interior, and eventually the initial awkwardness wore off. The rooms were still the same, but whereas before there was an accumulation of old furniture and other objects, now the space was mostly empty. I was especially struck by the absence of the floral sofa from which Nonna imposed her stillness on all the rest.

"I came to invite you to my wedding," I said when we returned to where we'd started.

"Your wedding? Really?"

"You're our neighbor, after all."

He said he would think about it, then corrected himself, adding that he was flattered and would certainly do his best to attend. I walked back through the courtyard. I hadn't talked to him about the gift, about the trip to an undecided destination, or about the amphora. So, in a way, my mission had failed. But Riccardo seemed so sincere, so grateful for my visit, that I hadn't felt like conning him.

Outside the gate I broke off two blades of grass and on the way home to the masseria I tried to braid them into a small crown, recalling the sequence of moves that Floriana and the boys had taught me. But I lost interest before I was able to do it.

BERN HAD ALSO correctly predicted the reaction of my parents. Though on the phone they hadn't had the presence of mind to show they were happy, they must have realized soon afterward that, once again, they could not stand in my way. My mother called back half an hour after the first conversation and was even a little moved.

"We'll go and pick out the dress together," she said, "don't even try to refuse. And I have no intention of buying it down there. Your father has already gone out to buy you a plane ticket."

Those words did nothing but reaffirm her contempt for the life I had chosen, for the place that she had detested long before I went to live there, but at that moment her voice and her peremptory tone were comforting. I remained silent so she wouldn't notice my weakness.

She added: "He can come too, if he'd like. But obviously he won't be able to see the dress. It's bad luck."

Oh, Mom, if you only knew how unlucky we already are! How it's bad luck that's forcing us to do this! My heart was thumping, longing to tell her everything. But one of the first decisions that Bern and I had made regarding the trip to Kiev was absolute secrecy. If even one single person besides us and Sanfelice were to know the truth, our daughter would always be only partly ours.

I didn't ask Bern to come with me to Turin. I was afraid that, if he did, I wouldn't be able to cope with his presence on top of that of my parents.

The fabric of the dress was so fine and delicate that when I tried it on I was careful not to touch it with my fingers, for fear of leaving fingerprints. It had an elegant crisscrossed cummerbund in front, whose bands joined to form a bow behind. Without a wrap, my back would be essentially bare.

The afternoon with my mother passed quickly, almost imperceptibly, as did supper back at home and the night spent in my childhood bed.

The dress would arrive in Speziale by mail within twenty days or so. If Bern were to ask how much it cost, I planned to say nothing and lie to him. The cost of the dress and shoes would have gotten us at least a thousand miles closer to Kiev.

A few weeks later I went with him to choose his suit. I'd had to convince him that he needed one; he insisted he could get by with what he had, Danco's clothes that he'd worn at my grandmother's funeral, possibly asking Tommaso if he could borrow something. I'd had to be as firm as I could, swearing that I would not marry him in a suit worn at a funeral, or in a waiter's uniform.

Inside the store, in a shopping mall outside Mesagne surrounded by industrial warehouses, he was as recalcitrant as a child. He'd take a jacket from the salesclerk's hands, scowl at the price tag, then shake his head and hand it back without trying it on. He went on like that until the young man ran out of options.

"You can't spend less for a wedding suit," I said, practically begging him.

"Two hundred euros!" he exploded, barely lowering his voice.

Suddenly I felt a great weariness. I slumped into a chair. Even with the air-conditioning, the heat was unbearable. The clerk brought me some water.

Seeing me like that, so pale, discouraged, and withdrawn, must have produced a reaction in Bern, because without saying a word he grabbed the two-hundred-euro blue suit from the rack, entered the dressing room, and came out a couple of minutes later with the pants dragging on the floor and the jacket open on his bare chest. When he spread his arms and twirled around, I could see his dark nipples.

He let the clerk bring him a white shirt, a pair of loafers, and a tie. The tie was garish, but I didn't say anything so as not to break the spell. Bern paid for everything and we left the shop, then the mall, and walked out into the parking lot that stretched as far as the eye could see, melting in the July sun.

HE AND DANCO got hold of some village festival lights, three impressive white arches with intricate forms in them: hundreds of little round lightbulbs screwed in one after the other. It required ropes to hoist them and props to hold them up. When they were all turned on together, they lit up the night at the masseria.

I didn't ask where they'd gotten them, just as I didn't question where the wooden tables and benches came from, or the tablecloths and dozens of candles, also white, to be hung in glass jars from the tree branches. Certainly Danco deserved much of the credit. He knew people throughout Puglia, people whom he could ask for favors.

The preparations swept me along to the day of the wedding almost before I knew it. I found myself rushed down the steps of Town Hall in Ostuni, clinging to Bern's arm, already his wife at that point, ducking between bursts of raw rice.

Afterward we walked down the dirt track to the masseria, as the sun faded slantwise and lengthened our fused shadows, Bern's and mine, from behind, so that they nearly touched those of the first row of fruit trees. The countryside and we two, one thing at last.

The guests followed us, grouped in small clusters; from time to time someone moved ahead to photograph us. Tommaso was the only one who'd stayed behind at the masseria, in order to supervise the young people from an agricultural cooperative who were acting as chefs and waiters.

Then night swallowed up the last remnants of light and we all found ourselves under the hundreds of tiny lit bulbs.

"There have never been so many people here," Cesare said, placing a hand on my cheek.

"I hope you don't mind."

"And why should I?"

"You always imagined it as a place of peace."

He slid his hand from my face to my neck, a touch so intimate that I would have drawn back with anyone else, but not with him. His presence on that day filled me with faith.

"I always imagined it as a sacred place," he corrected me, "and I can't think of a better way to celebrate it."

He smiled at me. He was searching for something hidden in my expression.

"Once I told you that in your previous incarnation you were an amphibian, do you remember?"

I remembered it, of course, but I was astonished that he did.

"Well, today I'm sure of it. You're able to adapt to many worlds, Teresa. You can breathe underwater and on land."

I was on the verge of confiding in him what was weighing on my heart even in the midst of that festivity.

We want to steal a baby. We want to steal our baby.

I sensed that he felt the presence of that secret. His eyes encouraged me, but I turned my head away.

"Thank you for coming," I said.

"Don't run off. I'd like to introduce you to someone."

I followed him under the pergola. Cesare touched the shoulder of a woman with loose black hair in a blue dress that revealed her slender legs.

"This is my sister, Marina. Bern's mother. I don't believe you've ever met."

But I had realized it before he said it, I could tell from her close-set eyes, identical to those of my husband. If I hadn't known the truth, I would have said she was his older sister. A child clung to her leg. Marina blushed.

"Bern told me not to bring him, but what could I do?"

"No, it's fine," I said, though I couldn't look at the little boy again, instantly reconstructing another portion of his life that Bern had kept hidden from me: the new family of the sister-mother whom he never mentioned, who had been added and then removed from the list of guests, and finally left there with a pen stroke that only partially crossed her out, present and absent at the same time. And a little half brother who would have been just a few years older than our daughter, if fate had been favorable to us from the beginning.

"Marina is very happy to meet you," Cesare said.

But she had already lowered her head to the child and whispered that he should behave.

"Had you been here before?" I asked her, just to say something. I remembered the piles of almonds, Bern's disappointed expectation, when his back had stiffened up from all the hulling.

Marina nodded. "I like the flowers you put in your hair," she said.

I wanted her to compliment me again. I hadn't known her until a few minutes ago and suddenly she was the most important guest at the party.

But she was finding it difficult, awkward. She said, "When are we leaving, Cesare?"

"After the cake," he replied amiably.

Then the child darted away, running between the forest of legs, as if fleeing from that conversation. Marina went after him, hurriedly apologizing to me. Cesare responded to my look with a hint of a smile, then he too turned away.

I TOOK PART in snippets of conversations, I laughed even when I didn't understand the jokes, I wandered around making sure that everyone was relaxed, that everyone had had enough to eat. From time to time I looked for Bern, and saw him surrounded by other guests, too far away. But I didn't allow that distance to bother me. I was determined to enjoy it all, every second of it.

Corinne snatched me away from a group of high school classmates who were asking me some leading questions about my life at the masseria.

"Your father is making a scene," she said, her face angry and tense.

"He's complaining that the wine is bad, and okay, so it's not good, but he shouldn't have attacked Tommaso. He accused him of serving it cold on purpose, to cover up the taste."

We reached the drinks table, where my father stood facing Tommaso. He took me by the shoulders.

"Here you are, good. We need to get some other wine, Teresa. This one is poison. Tamponi spat it out into the flowers."

Tamponi was his office manager. Even earlier, at Town Hall, my father's attention had been focused mainly on him.

"Don't we have anything else?" I asked Tommaso. He shook his head.

"But what were you thinking of, serving this stuff?"

"Maybe it's the bottle, Dad."

"I've already tried three. Three! And this guy keeps looking at me with that shitty smile!"

"You see?" Corinne burst out, as if it were my fault.

Tommaso said, "What do you want me to do, Mr. Gasparro? Oh, wait, I have an idea, bring me that amphora," he said, pointing to the amphora for the trip money. "Maybe I can manage to turn water into good wine. And if I can't, you can throw stones at me, like old times."

The most vivid part of my imagination saw my father lunge across the table to grab him, but luckily the musicians had arrived. Friends of Danco, enlisted from who knows where and in exchange for who knows what, they were his personal wedding gift to us (though Bern and I hoped he would not forgo slipping a little money into the amphora). Together the guests all moved toward the players and someone dragged me along and pushed me into the center of the circle that had formed.

The young man with the tambourine bowed to me and soon afterward Bern materialized, just as disoriented. He was the first to respond to the urging that came to us from all sides, moving his arms and legs as he whirled around me. He was better than me at dancing the pizzica, but what did it matter? I looked at him: my husband. I put my trust in him.

"Take off the shoes!" someone shouted. He bent down and untied my shoelaces, and I stepped onto the ground with my bare feet. That may have been the green light that the guests had been waiting for, because the circle around us broke and everyone started dancing.

Bern whispered in my ear that he was the happiest man on earth. Then, as if it weren't enough to confide it to me alone, he shouted: "I am the happiest man on earth!"

I lost sight of him and found myself dancing with other people, at one point even my father, whom someone had pushed into the fray. I danced for a long time, in a kind of daze. By the end my head was spinning. I was on the verge of stumbling, so I picked up my shoes, which everyone had carefully avoided, and made my way through the crowd to the pergola.

In the kitchen there were stacked pans, serving plates with leftovers, piles of dirty dishes. The young people from the cooperative were busily working amid that chaos, yet they smiled at me in turn, somewhat awed.

I went into the bathroom. The mirror returned the image of my loosened hairdo, the flower buds that Marina had complimented drooping to one side, my cheeks mottled. I was a little sorry to have lost my earlier composure, to see the unrefined farm woman I had become reemerge from beneath the makeup. I moistened a washcloth and used it to scrub my face.

At that moment the door swung open. In the mirror I saw Nicola towering there, he too disheveled, the knot of his tie undone. Instead of backing away, he closed the door behind him.

"I'll be out of here right away," I said, but he didn't move.

He was breathing hard. He took another step forward, grabbed me by the elbows, and lowered his face to the back of my neck, as if to sink his teeth in it. Raising his head, he started kissing my neck passionately, moving up to my ear, before I could break free. I banged my wrist on the edge of the sink, pushing him away.

"Get out!" I said, but even then Nicola didn't leave; wide-eyed, he was staring not at me, but at my image in the mirror.

"Get out, Nicola!"

He sat on the rim of the bathtub and looked around, as if making contact with the room again, with every single object. Then he covered his face with his hands.

I felt a little guilty now, seeing him so dejected, but I was afraid of what he might do if I got too close.

"What's wrong?"

He didn't answer.

"You had too much to drink. Why didn't you come with Stella? You could have brought her." He shook his head again. He got up, turned on the tap, and stood there blankly, watching the flow of water.

"Feelings are always so uncomplicated for you, aren't they?" he said tightly. "So clear-cut. But you still haven't understood a thing, Teresa. Not about me and not about this place. And not about the man you married, for sure."

I placed the damp washcloth on the sink so he could use it.

"See you outside, Nicola."

I opened the door. I glanced down the hall, in both directions, to make sure no one had seen us, that there were no witnesses to that betrayal that I had not taken part in.

THEN IT WAS TIME for the cake. I watched the creation pass by, carried by two young men; its rounded tiers were decorated with multicolored fruit, shimmering under a layer of aspic. The waiters took it to the holly oak, where a table had been set up. I didn't know that Bern had planned to serve it there. Once again I felt myself being dragged along reluctantly, and once again a circle formed around me.

Bern climbed up on the bench and held out his hand for me to join him. There were whistles and applause. Danco called for us to give a speech, and others joined him. But I wouldn't have been able to say a single word and Bern ducked his head behind my back. The guests quieted down, expecting one of us to say something.

That's when Cesare stepped forward: "Since the bride and groom are a bit overwhelmed, I'd like to say something in their place. If they give me their permission, of course."

I remember that moment very clearly, maybe better than any other: the trunk of the holly oak, Bern and I, the cake with the pieces of fruit arranged in circles, then Cesare, and farther on, the waiting crowd.

"Thank you, Cesare. Do save us," I said, before it might occur to Bern to stop him.

Cesare took another moment to collect his thoughts.

"Teresa and Bern chose not to join in marriage under the guidance of the Lord," he said finally. "This does not mean, however, that God is not looking down on us, on all of us, at this precise moment. Even though He hasn't been invited, He holds us in His warm, strong embrace. Can you feel it?"

He turned to the guests, his forefinger held up as if pointing to something in the sky.

"Can you feel that gentle pressure in the air? I feel it. It's the touch of His embrace."

I peered around at the faces of the guests with some apprehension, but only Danco had mockingly crossed his arms and was smiling scornfully. The others seemed truly captivated by Cesare, by his solemn pauses. I reached for Bern's hand. He was calm.

"I'd like to tell you a story," Cesare went on, "a story that you may not know. The story of the mutinous angels."

And so he spoke about the guardian angels, about their mutiny, about how they came down to earth attracted by the beauty of the women, of how they coupled with them and how from that union monstrous giants were born. About how afterward the giants rebelled against men and the earth was filled with blood and suffering. And about how the guardian angels taught men to defend themselves against what they themselves had generated, and taught them about spells and plant properties and how to make weapons. He spoke about all that to our guests, who were there to have a good time and maybe take a peek at our strange life, and they listened to him, whether out of curiosity or politeness.

Then he said: "I see that some of you have grown gloomy. Why is he telling us such a macabre story, you're wondering. Is he trying to ruin the party? What the heck is he trying to tell us?"

A few people chuckled and Cesare smiled too. He was full of passion now.

"That every glorious endeavor of man has its origins in transgression and sin, that's what. That every union between human beings is a union of light and darkness, even this marriage. Don't take offense, please. I've known our bride and groom since they were children, they are like a son and daughter to me. I know the transparency of their hearts. But the prophet Enoch would warn them against the darkness that exists within them as well, which perhaps they don't yet know. Teresa, Bern, remember this always: We marry virtue and sin at the same time. If you don't yet see it, dazzled as you are by passion, you will understand it later on. There always comes a time when it happens. And that is when you must remember your promise of tonight."

He looked around for Floriana. He stared at her for a moment, as if he were talking about them too, restating something important. Then he turned his back to the guests again and regarded only us, Bern and me, still standing on the bench, now a bit ridiculous on that podium.

"You were little more than children when you met, but perhaps you were already in love. Floriana and I talked about it. Isn't that so? Those two, we said, there's more than meets the eye there. Tonight you promised to watch over one another. Don't ever stop."

He took a few steps back to move out of the center. A few people clapped, but without conviction, and the applause died quickly.

In the midst of that hesitation, Bern stepped down from the bench, slid around the table, went over to Cesare, and leaned his head on his chest. When Cesare motioned me to join them, I too got down cautiously, and we both found ourselves in his arms, embraced by his blessing that we had so deeply missed, even if we hadn't known it until just then.

CORINNE AND TOMMASO were among the last to leave; he was so drunk and revved-up that they'd had to lead him to the car and put up with an angry outburst when he claimed the right to drive. When Bern and I were alone, we sat on the swing-chair, unconcerned that it might collapse under our weight. Husband and wife. Some of the ribbons that we had hung from the trees now lay on the ground, soiled.

On the table were some leftover wedding favors. I got up to get one and went back to the swing. I bit a sugared almond in two and offered half to Bern, but at that moment he started sobbing. I asked him what was wrong, but he was crying so hard that he couldn't answer. So I took his head in my hands.

"Stop it, please, you're scaring me."

His face was convulsed, splotched with red under his eyes, and he was short of breath.

"It was so beautiful . . . ," he stammered, "the most beautiful day of my life . . . everyone was here . . . did you see? Everyone."

He said it as if even then he had a premonition that nothing like that would happen again. And at that moment, for the first time, I understood the depth of his nostalgia, of how much he missed them all: his mother and his father, Cesare and Floriana, Tommaso and Danco, maybe even Nicola.

I stood up.

"Where are you going?" he asked, alarmed, as if I too might vanish.

"I'll make you a cup of tea."

"I don't want any."

"It will do you good."

Inside the house, I leaned my hands on the table. My dress was stained in the front, and it felt tight. I went into the bedroom and took it off, pulling on a pair of jeans and a T-shirt. I was about to leave the dress there, on the floor, but I spread it out on the bed.

Bern had calmed down, and was swaying gently on the swing-chair, staring straight ahead. He took the cup of tea and blew on it. I sat down where I'd been before.

We stayed like that for a while. Not a word about my having taken off the dress; maybe he hadn't noticed. He even seemed to have forgotten that he had cried nonstop for ten minutes, and about all the people who had flocked to the masseria, who, until a moment ago, he'd been sure he couldn't live without. Finally he stood up, lifted the amphora with the money, and hurled it against the concrete patio.

On our knees we separated the bills from the congratulatory notes and cards, the checks from the shards, opening the envelopes without even reading the good wishes. In the end, the table was half buried in bills; a gust of wind made them rustle and blew some to the ground.

We started counting. Cesare had never wanted money to be handled at the masseria, and now here Bern and I were, greedily passing it back and forth. If our guests only knew how different our first night as man and wife was from what they imagined! Under the white cotton tablecloth, Floriana's plastic one was still there, its map marred by ring-shaped burn marks, where piping-hot plates had been placed.

"Nine thousand three hundred and fifty," Bern said, after I'd handed him the last bills. He leaned toward me and finally kissed me. "We did it."

A hysterical joy grabbed us. All that money, and it was ours.

We went into the house. We took turns in the bathroom. Still damp from the shower, Bern climbed on top of me and entered me, clumsily shoving his way in, never taking his mouth off mine. Sex had become drained of meaning, ruined by the fear of failure and by the tons of hormones that had been injected, but not that September night. Although we moved more confidently than we had when we were seventeen, lying on the muddy ground of the reed bed, although there were no more surprises in the way Bern sucked my tongue, in the abruptness of my orgasm, or in the way he clenched his teeth surrendering to his, the unexpected frenzy of our bodies was a new revelation, and for a few seconds we didn't think about the future. For those moments of the night, only we two existed. But it was the last time.

A FEW DAYS later we brought some sugared almonds to Sanfelice. He chewed them one after the other, greedily. With his fingers still sticky with sugar he flipped through the pages of the calendar and announced that we would have to wait until January for the trip. October was not compatible with my menstrual cycle, November was booked, and in December he would be taking his wife and children on a week's skiing holiday.

He read the disappointment on our faces and to counter it he redoubled his enthusiasm. But January was perfect! There would be snow, mountains of snow over Kiev! Snow was auspicious, the success rate increased astonishingly. He left us sitting there, speechless, as he searched for statistics on the computer, then he turned the monitor toward us.

"Look here, February 2008. Pregnancies, one hundred percent."

Neither Bern nor I dared ask him whether snow was lucky in general, or only for him. Was there a reason, some scientific evidence? We were too dazed by fear and hope in equal measure. An unqualified one hundred percent, the doctor promised; he said the snow brought him luck, and we believed him. Having reached that point, we were willing to believe anything.

"Have you already arranged with the secretary for the hotel? We have excellent agreements. I stay at the Premier Palace, but some find it expensive. There's also a spa. A massage before the embryo transfer is good for you, it helps relax the tissues. For you, no," he said to Bern, "no massage, just a little elbow grease. Come on, you can do it! And let's hope it snows galore!"

I REMEMBER ALMOST nothing of the following months, only that I had another round of hormone therapy, somewhat different, less debilitating. The doctor's secretary phoned us about purchasing the airline tickets, she would take care of everything. So we were choosing the other hotel? Were we really sure? The price difference wasn't all that much and at the Premier Palace there would also be the doctor, which could be comforting. And would we be taking advantage of the guided tour of the city? From Tuesday (the day of the semen collection) to Saturday morning (the day of the embryo transfer) there wasn't a whole lot to do. The doctor recommended the tour to everyone, his clients often underestimated Kiev's attractions.

On New Year's Eve we went to Corinne and Tommaso's, but they weren't themselves, preoccupied by their little daughter, who had recently been hospitalized. They listened intently for the crackling of the baby monitor and took turns jumping up to check on her in the other room.

Danco monopolized the conversation, and when he finally fell silent, no one was able to fill the void. Giuliana yawned shamelessly even before midnight, and her sleepiness was contagious.

Bern and I got in the car right after the toast, grim and envious.

"They have a machine to chop the ice," he said. "Can you imagine how much electricity that thing consumes?"

But the time to pack our suitcases finally came and then the day of departure. At the airport Bern wandered around full of wonder. Everything was new to him. I had to practically lead him by the hand to the check-in desk and then to the lines for the security check.

He observed our luggage being carried along by the conveyor belt and then swallowed up. When we reached the gate, he called my attention to a Boeing maneuvering outside the window. As he watched it accelerate down the runway and peel off from the ground, he smiled like a little boy. Who takes the first plane ride of his life at age twenty-nine? I wondered.

I left him the window seat. He watched the foamy blanket of clouds for almost the entire time. "Imagine walking on it," he said, pointing out there.

To save money we had chosen a very inconvenient connection with a wait of almost nine hours in Frankfurt. Bern refused to set foot in one of the fast-food places because he was sure the meat came from intensive breeding, and the other restaurants were simply too expensive. We ate some chocolate first, then some bread miserably spread with mustard and pickles. By the time we finally got on the second plane, I was so hungry that I devoured the sandwich served by the stewardess, and immediately afterward the one she'd left on Bern's table as well, since he was asleep.

I tormented myself thinking about the cannula. A few days before Sanfelice had tried to insert it, "a trial run" he'd called it. It didn't hurt. It was just something that made you clutch the paper covering on which you were lying. "It's like an obstacle course with this cervix in the way," he had said. Then, victorious, he exclaimed, "There we are! We'll plant the damn thing right here!"

Would he manage to succeed as easily again? We had chosen to transfer the maximum number of embryos: three at once. So much the better if we had twins.

I WOKE UP with a start before landing. The food was churning in my intestines and I could still taste the mustard in my mouth.

"Are you ready?" Bern asked. He looked serious, as if he'd been deep in thought after waking up.

I suppressed the discomfort of my stomachache.

"Sure, ready."

Outside the Boryspil airport, the wind raised eddies of fine icy particles, sharp crystals that stuck to your face. My fingers were so numb that I could hardly put my gloves on. Our escort, Nastja, walked a few steps ahead of us, but unlike us, she did not bow her head to shield herself from the barrage.

"These are the warmest hours of the day," she said, with a rather militaristic Italian pronunciation. Her hair was tinted an unnatural red and was very short, with a single long strand hanging on one side. She laughed raucously. "Yesterday minus twenty. First time in Ukraine, huh?"

As if hypnotized, Bern moved off toward a rotunda in the middle of the parking lot, where the snow was about a dozen inches thick. There were no drifts as Sanfelice had promised, only a hardened layer of ice. Bern laid his bare hand on it.

"I didn't remember it," he said.

But I didn't feel like indulging his wonder at the snow, not while my legs and face were freezing, not while that woman was waiting for us at the car and my bowels felt like they were being twisted by a pair of forceps.

In the car, Nastja sat sideways in her seat to be able to talk to us. "You are worried," she said.

"No."

"Oh, yes, you are. You have fearful faces. Look here," she said, rummaging in her purse and pulling out a phone. She showed us a photo of two children. "Both with Dr. Fedecko. My husband, Taras, has a drunken sperm."

She puffed out her cheeks to mimic her husband's sperm. The children in the picture, radiant, were holding a tray full of bills.

"Thirteen hundred dollars. Won at casino," Nastja exclaimed. "Taras always wants boys, boys. Only boys. You have selected the sex?"

Outside the car window, a succession of gigantic buildings on the city's outer edge began. Was it girls living inside those concrete monoliths who offered their eggs in exchange for money?

Bern lit up with excitement at the sight of the frozen Dnieper.

Then, touching my wrist, he said, "Look! Look at that." On the hill was a group of golden spires.

"Pecer'ska Lavra," Nastja told us, "we go to visit it tomorrow. And that up there is the steel lady. Last Soviet statue, ordered by Nikita Khrushchev. See what big tits? Russian woman tits," she said, her hands making a vulgar gesture.

The pain in my belly had spread and now extended to my entire lower back. If I didn't get to a bathroom soon, there would be a disaster.

"What's wrong?" Bern asked.

"How much longer to the hotel?"

Nastja pointed vaguely ahead. "After bridge comes center."

Then she said to herself: "Faces very, very fearful, yes."

The columns in the hotel's lobby were overlaid with a plastic coating that imitated veins of marble. The red carpeting ranged everywhere. The male staff, all in livery, were sitting in the corners with a drowsy air;

their eyes followed us as we handed over our passports, filled out the registration forms, and got some final instructions from Nastja.

"Down here at five o'clock, with a nice jar of sperm. Nastja's advice: a glass of vodka first, just one, and a slice of salo, lard. The lard makes the semen stronger. Secret of Taras."

We rolled our suitcases toward the elevators. I had the impression that everyone knew why we were there.

The room on the second floor had a single window overlooking a parking lot full of debris. Across from it was a building gutted by a collapse or maybe never completed.

I locked myself in the bathroom while Bern sprawled out on the chenille bedspread. I filled the tub and remained soaking in it until I melted, even though the water coming out of the pipes seemed to be as contaminated as everything else. But at least it was piping-hot and helped relieve my chills.

Bern took Nastja at her word. I wanted to stay in the room, crawl under the covers, and wait, but he made me get dressed. We had to go look for the salo.

Outside, on Khreschatyk Street, columns of cold, harsh Siberian air advanced on us. We walked for more than half an hour, first along a park, then downhill on the avenue that led to the train station. The plaza in front was a huge, treacherous slab of ice, and the humanity that populated it, males only, with caps lowered over their eyes, made me beg Bern to leave there immediately.

We went back up the same street and ducked into a café that seemed stuck in another century: lace curtains on the windows, wood-paneled walls, blinking Christmas lights. Bern managed to order the lard. The woman brought it to him cut in thick strips, with pickles on the side.

"It looks disgusting," I said.

"It's for the cause," he replied, amused, then he picked up a slice of fat between his thumb and forefinger and dropped it into his mouth. My bowels roiled again. Bern ate all the salo on his plate.

There was still time; he wanted to walk. It was the excitement of his first trip abroad, so far from Speziale, except for the mysterious year he'd spent with his father in Germany, too young then to remember it except in confused flashes, and which he never spoke about in any case. Even the clouds of condensation that came from our mouths seemed extraordinary to him. I decided to let his exuberance rub off on me. It was our honeymoon trip after all, bizarre, troubling, but still our honeymoon. As far as Danco and the others were concerned, at that moment we were in Budapest, being tourists. I could at least pretend it was so.

When we returned to the hotel the other couples were gathered around Nastja in the lobby lounge. The woman spread her arms out to us and in an embarrassingly loud voice exclaimed: "Here they are, the two missing ones. Jar, quickly!"

She asked Bern if he needed magazines or photographs, she had plenty in her bag. He refused, though fascinated by such boldness. He asked me to wait for him there.

Nastja led me to the armchairs, practically forcing me into the only vacant one. The lady beside me turned to me and said, "Yesterday my endometrial thickness was fourteen millimeters. Sanfelice says it's perfect."

She didn't say her name, didn't offer her hand, didn't choose words generally used to start a conversation, she simply told me the thickness of her endometrium, then added: "It's the seventh time we've come. But it was always thinner. Besides, did you see how much snow there is on the streets?"

I kept staring at the closed doors of the elevator at the end of the hall, until Bern reappeared. He crossed the empty lobby and handed the sample to Nastja, in front of everyone, without a trace of embarrassment.

"The latecomer," she said, then studied the jar against the light. "Good, good, there is a lot. Do you know what they say here in Kiev? That you must always stock up for a dark day. Because sooner or later it comes. It always comes. Cherniy den', the dark day."

A BLIZZARD KEPT us confined to the room for two days. The gusts were so violent they made the windowpanes shudder. The wind was called buran, the driving snow was purga. Bern found it amusing and kept repeating it: buran, purga, purga, buran.

I couldn't bring myself to do anything. I lay on the bed, staring at the damp stains on the wallpaper, trying to guess its original color. Lying next to me, Bern studied the travel guide. Occasionally he'd read something aloud, then he'd look for a pencil and underline the passage that interested him.

But on the third day, the day before the transfer, the sun was shining, dazzling because of the snow, though it provided no warmth. Nastja was waiting for us in the lobby, for the guided city tour. I didn't want to go, but Bern couldn't see why: we were there, we had the whole city at our disposal, and it was such a radiant day.

"Brave souls," Nastja said, spotting us. "Let's go."

The city seemed hostile and frightening to me, just as at first: the underpasses with their stifling shops, the homeless prostrate from alcohol, then down into the subway, on escalators so long and steep they seemed to lead to the bowels of the earth, the names of the stops written in that incomprehensible alphabet. Bern and Nastja were always a few steps ahead of me, closely engaged in a conversation that I had neither the strength nor the desire to follow. A suffocating heat inside the buildings, a paralyzing cold outside; I tried to cover my mouth and my nose with my scarf.

On the Andriyivskyy Descent, I lost my balance twice. Bern turned to look at me with a strange indifference, almost as if he were annoyed. He was attracted by the stalls and insisted on buying a gas mask from the Cold War era, which Nastja helped him put on.

"Danco would like it," he said. But we were short on money and we weren't sure if anything like it could be found in Budapest, so he left it there.

I looked at the young women around us. They were as beautiful as Sanfelice had promised, tall and slender, with dark hair and very pale complexions. It could be her, I told myself, meeting the clear gaze of a passerby. What will her name be? Natalija? Solomija? Ljudmyla? And will she have other children? I couldn't stop those thoughts and I didn't dare confide them to Bern. He would have told me to stop being silly, he would have quoted Sanfelice's words the way he used to quote the psalms.

I persuaded him to take a taxi to the hotel. Nastja backed me, I had to be rested for the following day.

As we went up the wide, tree-lined avenue, the car radio played a song I knew. I hummed a few bars in a low voice.

"What is it?" Bern asked.

"Roxette. 'Joyride.' I used to listen to it when I was a kid."

The driver caught something, because he said, "Roxette, yeah! You like music nineties?"

I said yes, I liked it, but mainly not to dampen his enthusiasm.

"I also," he said, his limpid eyes looking at me in the rearview mirror. "Listen."

For the rest of the drive, all the way back to Khreschatyk Street, he chose one song after another for me, each time looking for me to confirm that I liked it: "Don't Speak," then "Killing Me Softly," then "Wonderwall." I looked out the window; the sun had set quite a while ago, the areas of the sidewalk away from the streetlamps remained in the dark.

In front of the hotel entrance, Nastja said, "Tomorrow money, remember. In euro cash."

A STORK PERCHED above the doors of the clinic. It was stone, but so well carved that at first I mistook it for a real one. Sanfelice had chosen to proceed in alphabetical order according to the surnames of the women, and we were the third couple.

We found ourselves in a modern space, a taste of the future in the middle of a neighborhood where everything else looked old and worn. Nastja held me by the arm, as if I might escape. Before I stepped past the doormat, she handed me two plastic bags.

"For shoes. You, too," she told Bern who was following us in silence. She was treating him more brusquely than she had the day before, as if at that point he were only an impediment.

I covered my shoes with the blue plastic. That sanitary precaution should have reassured me, but instead my tension mounted progressively as I went up the shiny polished stairs, as Bern was diverted to another corridor with no time for us to say goodbye, and as I filled out the preprinted forms, written in an English full of grammatical errors, requiring my consent for freezing the embryos and for disposing of them after ten years.

Then I was lying in an operating room equipped with machinery and lamps, the walls covered with tiles up to the ceiling. On one side was Dr. Fedecko, extraordinarily tall, with a blond mustache, and on the other side Sanfelice, with his usual cheerful air, though more subdued than usual.

"We have competitive-quality blastocysts," he said, "all 3AA. Those of the lady before were only B, just for comparison."

In the meantime, Fedecko advanced with the cannula, looking for the least resistant path to enter me, more delicate than Sanfelice had been during the trial run. It was over in an instant. The doctors complimented me, though I wasn't sure for what. I had simply lain still, nothing more, and everything that had ensued didn't seem to really concern me, or it concerned me only indirectly.

I was brought to another room, smaller, with a large window. I waited for what seemed like a very long time. I could see the snow-covered hill, and in the center of that whiteness the golden spires of Pecer'ska Lavra. We had visited it the day before, but it was more enthralling seen like this, from afar, like a mirage.

I felt cold. And where was Bern? Suddenly I convinced myself that he was no longer in the building, maybe not even in the city, and everything became distant, unreachable, like the miniature of the Lavra on the hill.

Then the door swung open. Sanfelice and Dr. Fedecko entered along with two nurses and, behind them, Bern. He didn't dare approach the bed, except after we were left to ourselves. Then he helped me get up and put on the clothes that a supernatural hand had transferred from the room where I had taken them off to the closet in this room.

With no one to escort us anymore, we made our way alone through a maze of corridors. We went down more stairs and found ourselves in the lobby. Nastja was there. She bent down to remove the nylon shoe covers I again had on my feet, and pointed to the car waiting for us outside.

THE VEGETATION at the masseria was dozing thanks to winter. Bern and I were as suspended as the nature around us. He studied me silently, looking for any change in my body, in my metabolism, in my sleep. I quarreled with him about the littlest things—for example, over the fact that he hadn't swept the pavement in the yard and the leaves had blocked the drain. Actually, I wanted to shout at him to stop following me around, stop asking me how I felt, stop boring into me with his eyes wherever I went! The truth was that I felt exactly as I had before, only more listless, more irritable.

So I really wasn't surprised when Sanfelice, after circling the intrauterine probe around and consulting the monitor full of jumbled shadows, announced that there was nothing, nothing that was moving.

"What a shame. All those fantastic blastocysts. However, the next trip is in March."

Bern had not accompanied me to the visit. "Let's act as if it's a day like any other day," he'd said.

When I phoned him, he was at the farmers' market in Martina Franca. He left me on hold while he finished serving a customer. I listened to the exchange between them, then I pictured him crouching down, hiding under the table to gain some privacy. Conspiring had become a habit for both of us.

"Well?" he said in a low voice.

I told him the results without preamble, almost brutally. Then, immediately remorseful, I added: "I'm very sorry for you."

"It's all right," he said, but he was breathing hard.

"I'm sorry for you," I repeated.

"Why are you saying that? Why do you feel sorry for me?"

"I only just realized it. But it's true. I feel sorrier for you than for myself."

"You don't think that, Teresa. You're just upset. You don't really think that."

"You should find another woman, Bern. One who functions better."

It was during the ensuing interval of silence that I realized I was right. Before Bern could say it wasn't true, that I shouldn't talk that way, that I was only being silly. A very brief pause, no more than a split second's hesitation, the time needed to take a slightly deeper breath. He was considering the possibility that I had offered him. For a moment he weighed and compared two impossible choices: his desire for me and his heart-wrenching yearning for a child. This too could happen. It could be that in life, irreconcilable desires develop in people. It wasn't fair but it couldn't be avoided, and it had happened to us.

His uncertainty told me which of the two desires had prevailed, though he now denied it as forcefully as a telephone conversation in the middle of the market allowed him to. But I wasn't angry with him. On the contrary, I felt calm, as lucid as I'd been the night of the cramps. In fact, I didn't feel anything anymore.

I said: "Maybe you don't realize it now. But in five years, or ten or twenty, it doesn't matter how long, sooner or later you'll realize what I took from you and you will hate me for it. For having ruined your life."

"You're talking nonsense, Teresa. It's the disappointment talking for you. Go home now. Go home and rest. We'll make another trip, another attempt."

"No, Bern. There won't be another trip. We've gone far enough. And it wouldn't do any good. Don't ask me how I know, but I know."

I could see him, Bern, more and more huddled under the table, the din of the market around him.

"We're married, Teresa."

He affirmed it sternly, as if that were enough to put an end to the discussion. It wouldn't have worked, not like this. Bern would have insisted, pleaded if necessary, then we would have gone home and he would have patched things up with his ready phrases, one by one. The gleam of his black eyes would have won me over in the end, and we would have started all over again. Another absurd hunt for money, another round of treatments, another trip in vain to the most inhospitable place on earth, then the disappointment and so on, ad infinitum, until it destroyed us both.

I recalled the expressionless face of the woman in the armchair at the hotel in Kiev, the doggedness that had transformed her year after year. I did not want to end up that way. We were still young.

I said: "We made a mistake."

"Stop that!"

Strange, the reversal of roles between us, I hadn't foreseen it. I hadn't foreseen any of this. From the beginning I had been the one prepared to be abandoned; I had loved him from a distance, miles away, like a fool, while he had gotten someone else in trouble. And maybe it was because of the repressed memory of that summer, left unspoken, that I now knew what to do, how to halt the spiral into which we'd been driven, a spiral triggered the moment when Tommaso and Corinne had told us about their baby and we had begun to fantasize about ours. Yes, there was only one way to give Bern back his freedom and regain mine.

"There's someone else," I said.

"Someone else?" he repeated, practically whispering.

I knew him well enough to know that this was the only way to go. I was lucid and self-possessed. I was worn out and full of rage and my heart was broken. I would not stop.

"Yes. Another man. For me."

"You're lying."

I stopped answering, because if I had, he would have known.

Then his voice underwent a transformation. For a few seconds Bern became someone else, someone I had never known, someone furious.

"It's him, isn't it? Is it him, Teresa? Tell me, is it him?" he shouted.

"It doesn't matter who it is."

Those were the last words we spoke to each other, for a long time. "It doesn't matter who it is." In fact, those were almost the final words of our brief, unfortunate, and nonetheless indelible marriage.

I DID NOT go back to the masseria. I drove around aimlessly until after sunset. Later I would not be able to reconstruct my route through the outskirts of Francavilla, then through the maze of dirt country roads that sometimes ended abruptly at a gate, with guard dogs rushing to the fence, barking as though possessed.

I went back to Speziale, but I could not go home. I had a feeling that Bern was there, waiting to see with his own eyes if what I had told him on the phone was true, waiting to interrogate my body instead of my voice.

Only by spending the night away would the nonexistent betrayal that I had confessed assume solidity.

It's him, isn't it?

Before reaching the dirt track to the masseria, I turned into my grandmother's villa. I rang and waited for the electric gate to swing open, the flashing light at the top intermittently revealing the countryside.

Riccardo came toward me, wearing a tracksuit. I asked him if I could stay there for the night, in the lodge. It was a brazen, almost ridiculous request, but I must have looked so distraught that he said: "Of course, but it's freezing cold in the lodge."

"That's all right."

"The guest room is free. Come inside. I'll go get you some sheets."

The guest room was my old room. After he handed me some sheets and towels, and after I refused his offer to eat something, Riccardo closed the door, wishing me good night, sensing that in another moment his presence would have been intolerable to me.

And so, there I was back where I had started, in the room of my childhood, the villa dark for hours by now and me still awake, no sign of sleep, only exhaustion, that edgy exhaustion that has already ruled out any possibility of sleep. Lying in the bed where it had all begun.

A glow filtered through the shutters. The rising moon, I thought. But it couldn't be the moon, because it was a flickering light. I got out of bed and threw open the window, letting the cold air hit me, and I saw the blaze, the changing glimmers of the flames and the column of smoke rising straight up in the absence of wind before dissolving in the black sky, right where the masseria was. Neither the sound nor the smell of the fire reached me, there was only that slash of light among the treetops.

My first instinct was to rush over there, but it took only a moment to realize that it was only a signal, an ultimate appeal launched by Bern against the night, so that I would run to where he was and take back what I had said on the phone. A pyre to say: As long as it burns, I will wait here, ready to believe whatever you say, ready to forget. But when the flames have died and the embers are cold, I won't be here anymore, and what you said will be true forever.

I wondered what he had set fire to, whether it was the tool shed, the greenhouse, or the house itself with everything that it contained, mine and his. The next day I would discover that he had burned the woodpile, the entire reserve of firewood. But in my old room at the villa I didn't know it yet. At that moment I could only go on watching, my feet rooted to the icy stone floor; I did not reach for the blanket on the bed and wrap it around my shoulders, I just watched, until the flames diminished and finally, at dawn, died out completely.

TWO DAYS after our abrupt and shocking separation, I returned to Turin. I felt that my time at Speziale had ended. Now that I was too old, I would continue everything that I had abandoned when I was still too young.

I didn't even stick it out a month, however. The cold, efficient ways of the city, the rain followed by the luminous, heart-wrenching days of March, but above all my parents' wary forbearance, their tacit satisfaction over the failure, left my nerves constantly on edge. By then my life was in Puglia, so I went back. Not with the trepidation I'd had when I was a girl or with the relief of the last few years, but with a tenuous resignation, as if by that time I had no alternative. I was certain that I would not find Bern there, and in fact I didn't.

At times I didn't sleep because I was afraid. My head was filled with the macabre stories that I had heard about that countryside over time: a man had been attacked in his home, hands and feet tied, and tortured for hours with a red-hot iron. They were nothing more than old wives' tales of course, but in the darkness and the silence I let myself imagine things. One night I heard something metallic slamming about outside, very close to the house. I opened the door, trembling. A dog had his snout in the overturned garbage pail and was rummaging through it; he stared at me for a few seconds before trotting away.

But eventually I got used to it. After Danco and the others left, I felt it had all been a slow preparation for solitude, which was now complete. I accepted the simple comfort that nature was able to offer me. To lessen the feeling of loneliness at least a little, I bought a goat, which I let wander freely around the property. I started going to town more often and joined the parish choir and an amateur volleyball team at the recreational club. At the masseria, I had a phone line and an internet connection put in. The company's technician, a young man with long hair tied back in a ponytail, walked around the grounds holding up a metal pole in an attempt to link to the best cell, as if it were a divining rod. He set up the antenna and was amazed at my utter incompetence where computers were concerned. He gave me the necessary instructions and his business card, just in case.

One of my grandmother's former students, who now taught at the elementary school, came up with the idea of arranging guided tours of the masseria. The project that we had initiated there was vital, she said, and I could convey a respect for tradition and for the land. I was skeptical at first, since I had no experience with classes and I did not feel authorized to talk about the principles that we applied at the masseria; they were Bern's and Danco's, I had merely imitated them. But it turned out to be easier than I thought. I found myself teaching how mulching enabled us to save as much as ninety percent of the water, and why it was therefore crucial. I explained to the children why a spiral vegetable garden was more efficient than the rectangular ones they were used to and made up contests in which they had to recognize the aromatic herbs blindfolded, by touching or sniffing them. I let them sow seeds and water the plants, and when I wanted to stir them up, I showed them how the compost toilet worked, how it served to fertilize the soil.

As for Bern, I knew that he had been drifting for a while, but that he was now living in an apartment in Taranto with Tommaso, after Tommaso and Corinne split up. It was Danco who told me. I didn't hear from any of them anymore, but he showed up at the masseria one day, sent by Bern himself, with a list of things to take.

"He could have come himself," I blurted.

"After what you did to him?"

Perhaps he realized how tactless he'd been, because he added: "Anyway, it's none of my business."

He moved brazenly through the rooms, as if the place still belonged to him. He consulted the sheet of paper with Bern's handwriting.

"How is he?" I asked.

"He's doing okay."

Knowing he was all right should have comforted me, but I wasn't capable of such generosity. I sat down at the kitchen table, suddenly weary, watching Danco rummage through the drawers.

"Bern is made for lofty ambitions," he said at one point. "None of us has the right to restrict him."

"Is that what you think I did? Restrict him?"

Danco shrugged. "I'm just saying that before you showed up at the masseria we had plans. And now we can resume them."

"And what plans would they be? I'm curious. Liberating the cows? The sheep?"

He turned to look at me. "There is something more important than ourselves, Teresa. You've always been a slave to your idea of happiness."

But I wasn't willing to put up with his lectures, not anymore.

"And these plans of yours, do you realize them with the money from my grandmother's house? Don't touch that coffeepot, put it down! I was the one who bought it, it's mine. If Bern put it on the list, he was mistaken."

He put it back. "Whatever you say."

I waited for him to complete his mission. I sat there the whole time, filled with a grudging bitterness.

Before leaving, Danco waved to me. On the table under the pergola, I found the sheet with the list: written on the back was Tommaso's new address.

I DIDN'T HEAR anything more about Bern for a year. Until the morning I was awakened by the crunch of tires on the dirt track; it was not long after daybreak.

I reached the door a fraction of a second after someone started knocking firmly. I didn't ask who it was before opening; I grabbed my parka from the coat rack and put it on over my nightgown.

One of the policemen introduced himself, but I wouldn't remember his name, maybe I didn't even catch it. He said, "Are you Mrs. Corianò?"

"Yes."

"The wife of Bernardo Corianò?"

I nodded again, though it was strange to hear Bern mentioned in the chill of dawn.

"Is your husband at home?"

"He doesn't live here anymore."

"You haven't seen him today?"

"I told you, he doesn't live here anymore."

"And do you have any idea where he might be at this time?"

Something prompted me to say no, a vague protective instinct. Somewhere I still had the sheet with the address left by Danco, and in any case I had memorized it from looking at it so often. But I said no.

You promised to watch over one another . . . Don't ever stop.

"Would you rather we came inside and sat down, ma'am?"

"No. I'd rather we remain standing. Right here."

"As you prefer. I suppose you aren't aware of what happened last night," the policeman said, stroking his chin, as if embarrassed. "It appears your husband is involved in a murder."

"You must be mistaken," I said. A nervous laugh escaped me.

"There was a confrontation over cutting down some olive trees. He was among the protesters."

The light on the countryside was whitish, opaque.

"What murder?" I asked.

"A police officer. His name was Nicola Belpanno."

5.

Tommaso's hands still rested on the bedspread. He looked at them without lowering his chin, moving only his eyes, as if he were seeing through them to complete the fabric's geometric pattern of red and blue diamonds. The fingers were spread as if to say: That's all, there's nothing more to know, and this time I didn't leave anything out.

So there was the story I knew and then there was another, secret one. In that secret one, a girl and her baby had died. But Bern had never spoken to me about any of this, he had kept the promise he'd made to the others until the end. Not one story, but two, I kept telling myself, both real, as real as me and Tommaso in the flesh, in this room where the radiators had been turned off for hours. Two versions like the opposite edges of a box, impossible to see together, except with the imagination. The imagination that I had stubbornly refused to use where Bern and Violalibera, their child, and the other boys were concerned. Blind and deaf, and something even more grave. Obstinate. Unshakable.

Yet I said nothing. I hadn't even said: So that's how it went. I'd been silent since Tommaso had described Violalibera tied to the olive tree. And now he was silent too. Five minutes had already gone by like that, maybe more.

Then he said, "Could you check on Ada?" and I was almost relieved to get up.

I approached the sofa, until I could make out the slow, rhythmic rise and fall of the blanket covering Ada's chest. I gave it time to calm me down, then I returned to Tommaso, unsure whether to sit in that torture chair again or remain standing up.

"She's sleeping," I said.

He had moved his hands, those bloodless hands of an eternal child, and folded them on the turned-down sheet.

"I have another favor to ask," he said. "Medea should be taken out."

I looked at the dog curled up at the bottom of the bed, maybe on Tommaso's feet.

"She seems to be snoozing blissfully."

"I'll go, I can do it."

Flinging the blankets aside, he set one foot on the floor. He was wearing only white boxers under his T-shirt. The sudden sight of his naked legs threw me for a moment. He stood up but lost his balance almost immediately.

"Maybe I'd better not," he said, lying down again. "As soon as I move, everything starts spinning again."

Reluctantly, I took the leash that was hanging from the nightstand. Medea stood up at the faint clink of the snap hook. She jumped off the bed and barked twice before Tommaso commanded her to be quiet.

"If you see other dogs, hold on to her as tightly as you can. Even if they're behind a fence. She can make impressive leaps."

I HEADED toward the port. Medea moseyed around, sniffing every inch of the sidewalk for invisible traces of other dogs. It was the weirdest Christmas I had ever spent in my life.

What if Violalibera had wanted to keep the baby? If she hadn't been alone that morning, if the first sip of the oleander tea had disgusted her to the point of pouring the rest in the sink? It was strange to think that your destiny depended on someone else's decision, on that person's moment of weakness. As disappointing as being deceived. "In thought, word, deed, or omission," the catechism said, but nobody ever worried about the omissions. Bern and I hadn't worried about them either.

And yet, that was the first time I hadn't been lonely in months: as I walked to the port with Medea, without another human being around. As if now that I knew the facts, my life stretched backward and sideways, in every direction, crossing over into that of Violalibera and Bern and the other boys. As if I had finally dived into the pool with them, where they swam furtively that first night. Bern would have been able to express those thoughts better than me.

I looked at the dark blob that Medea had left in the middle of the sidewalk, then bent down and used one of the bags that were tied to the leash.

TOMMASO WAS DOZING sitting up. As he'd explained to me at the beginning of that night, it was the only position that didn't cause the furniture in the room to crash down on him as soon as he closed his eyes. I touched his arm gently, but he didn't wake up. So I shook him a little more forcefully.

"Wha-a-t?" he groaned.

"So you weren't sure."

"You know, even in Guantánamo they wouldn't force someone in my condition to stay awake."

"You weren't sure whose it was."

"Each of us was convinced that it was his and each of us was convinced that it wasn't. I don't think it can be explained any better than that."

"And you decided to wager the paternity with stones."

Tommaso didn't move. These were things that had already been said; by repeating them I was merely being cruel.

"But Bern chose to lose on purpose," I continued. "He wanted the baby for himself."

Or her. But neither Tommaso nor I said so.

Medea was snuggling at the foot of the bed again, as if she'd never left there.

"And Violalibera didn't say anything? Didn't she have a right to express a preference?"

"Bern had talked to her earlier. I think. He must have."

"Maybe any one of you would have been fine with her. What if it had been you?"

Tommaso turned his head to face me. He had never looked at me so deliberately as he spoke, it took me by surprise. Then he turned back to the bedspread, slowly, maybe because the sudden movement of his head had caused a stabbing pain.

"I imagine he had explained his intention to Violalibera, that he had promised her he would make the shortest throw. They had started together and they would finish together. A pact between them, something like that. I don't know, I didn't think too much about it at the time. Now I think that Violalibera may have realized later that none of us was really fine with her, on the contrary. She was a very strange girl."

Tommaso rubbed his face, then pressed his palms over his eyes.

"I want to know about the encampment, about that night," I said.

"I wasn't there."

"But this is where Bern was living, right? Here with you. They came to look for him at the masseria, but he was here before he went to that place."

"It's two a.m., Teresa."

But I didn't budge and Tommaso saw that I would not let up. So after a somewhat longer silence, he surrendered: "Okay. Go get some wine, though. There should be an open bottle under the sink. Unless I've forgotten that I finished it."

"Are you serious?"

"It will do me good. I'm a pro, I told you. And besides, is it or isn't it Christmas Eve?"

I found the wine, poured him some, and went back to the bedroom.

"Remember the day of the bees?"

"What do bees have to do with anything?"

"Corinne had told me the night before. About the pregnancy. About Ada, I mean. It's always been a habit of hers to let me in on a situation after the fact. Like the first day at the masseria, when she took my head and pressed her mouth against mine in front of the others. As if to say, Now I'm your girlfriend, they've all seen it."

"I thought you were in love with Corinne."

Tommaso took a deep breath.

"I suppose, for some, certain questions are more complicated than that. Anyway, Corinne decided to stop taking the pill. Without asking for my okay, with her usual blind, self-centered determination. But the baby, Ada, wasn't her real objective. She had never dreamed of having children, she didn't care about them. A pregnancy was simply the quickest way to drag me away from the masseria once and for all. I sound unfair, I know. She must have said horrible things to you about me in recent times."

"I haven't seen her in ages."

"But she didn't do it consciously. The truth is that she hated the masseria. She'd been okay with it as long as it meant spiting her father, but, once that was done, she had seen it for what it really was: harsh, squalid, grueling. I don't mean to offend you."

He'd begun talking somewhat guardedly, but now it seemed he wouldn't stop.

"Then too, Corinne couldn't stand Danco anymore: his arrogance, his tirades. But she knew I wouldn't leave there just because she asked me to. Even though by then I considered her my fiancée for all intents and purposes and I never, or at least rarely, thought about how it had started, especially after you arrived."

"Why me?"

With his thumbnail Tommaso followed the outline of one of the bedspread's diamonds.

"We were all paired off at that point, right? But Corinne knew I wouldn't agree if she asked me to leave. So she didn't even try, she went off the pill instead, for one or two weeks, for months, who knows. As long as it took. And even when she was already five, then six days late, she didn't say anything to me. This too I would reconstruct later on. She didn't say a word even after the pregnancy test, when she was sure. She talked about it with her father and mother first and let them take her to the gynecologist: the three of them together, the family reunited. They also chose the apartment that would become ours. Then she announced to me that she was pregnant. She looked only slightly guilty, but otherwise ecstatic and triumphant. She said that we would leave the masseria as quickly as possible, that the apartment, on the top floor, would be ready in a few weeks, and that all we needed was some furniture, which she wanted to choose with me. Then she said: 'You don't have to worry, my father has taken care of everything,' and with those words she obliged me to forget all the horrible things she had said to me about him. That night I thought I could already hear the breathing of the creature in Corinne's belly. I kept telling myself: You have to fix everything in your memory, because this is the first of the final nights."

Tommaso was more and more focused as he spoke, while I was thinking: This is how life chooses where to grow. It chooses without choosing, it germinates in one place and not in another, at random. Corinne and Tommaso and their dysfunctional love were fine. Bern and I weren't.

"The following day was Sunday," he continued. "Maybe leaving the masseria at dawn to go to work would hold off my despair, would keep me from staying in bed with Corinne and all those thoughts. The idea of sitting under the pergola, with all of you, aware of the short time I had left there, terrified me. So I jumped out of bed, grabbed my clothes, and went out. I wandered around for a while, before finding myself at the reed bed. The sun's rays filtered through the leaves. I saw the beehives. Really I hadn't thought about it beforehand. I didn't seriously think about it even as I lifted one of the lids, hypnotized by the turmoil inside, by that sticky, teeming mass. The bees weren't frightened, they just seemed a little agitated, as if startled by the shadow of a cloud. I put one hand in cautiously, then the other. They latched onto my fingers and wrists, searching for something. Abruptly I closed both fists. I don't remember anything else, only Bern, sitting beside my bed, in the hospital, a little like you're sitting there now, only on the other side, because I had to turn my head to the right to look at him. My whole body was throbbing, but there was no pain, and Bern's image was out of focus because the flesh had swelled up on my cheekbones and eyelids. I tried to say something to him, but my tongue was numb and he ordered me to be still and not talk. He promised he wouldn't leave while I slept.

I didn't want anyone else, only him. I hope my saying that doesn't bother you."

Did it bother me? Was I jealous as I listened to him talk? Maybe not. For the first time, maybe not. How silly, that rivalry between us. As if everyone's heart had room for only one person, and no more. As if Bern's heart hadn't been a convoluted rabbit warren, full of tortuous burrows, one for each of us.

"Go on," I said.

"The apartment had so many closets that Corinne's and my clothes didn't even fill half of them. For a month we did nothing but buy. She'd wait for me to come back from the Relais, then we'd walk downtown, exploring the shops. Items for the baby mostly, but also clothes for her and for me, and household appliances, because the kitchen was also half empty: a blender, a toaster, a yogurt machine, and even a popcorn maker. Corinne paid for everything, with a brand-new credit card. We were so different than we were before, unrecognizable. And we never talked about the masseria, or about all of you. I wasn't unhappy, not exactly. There was something liberating about being away from it, being rid of all of Danco's restrictions. I liked seeing Corinne radiant, playful, in a way she was never able to be at the masseria.

"We chose the baby's name and we got used to referring to her that way. Day by day she became more and more real . . . But no. That's not entirely true," Tommaso corrected himself. "Dismembered, that's how I felt. Dismembered."

It annoyed me that he kept losing himself in these digressions. It was exhaustion, that and all the alcohol.

"Because I belonged to Corinne and to Bern," he added, then burst out laughing.

"You'll wake Ada that way."

"Or, rather, no," he corrected himself again, still laughing loudly. "I belonged to Bern, period. That's really what I mean. But I was very confused at the time. Does it upset you to hear me say it? You have every right to be upset."

He rubbed his forehead, as if to make room for different thoughts.

"In the morning Corinne would be lying beside me, and I'd again tell myself: Stop thinking, follow the series of everyday movements that will begin in a moment, and you'll see, it will be better. And on and on for the rest of your life, every day, from here on out. So that . . . well, I was counting the weeks, as I lay with my eyes wide open beside Corinne, who was pregnant. I counted the weeks remaining before the birth, and when there were five I told myself five more and then I'll have to find another way. You see what I'm talking about? Sex, that's what. It would all have gone well enough if it hadn't been for that one little detail, sex. But it's not such an insignificant detail for a couple, right? No, it isn't. And you know something else? I spent a lot of time imagining how it was between you and Bern. It's horrible, I know. But by now this is where we're at. The whole truth and nothing but the truth, Teresa. I imagined how it was between you and Bern. No morbid details, that wasn't it, even though I occasionally gave in to those as well. More than anything it was the feeling that I was missing, what it felt like to surrender to such a blissful, complete attraction. So I counted the weeks before that truce ended. Because I could love Corinne immensely, but I could only love her minus the sex. Assuming it means something. She knew it too, I think, even from our time at the masseria, but she was convinced she could change me, that she could correct me. And if she couldn't, habitual practice at least would straighten me out. Generally Corinne was determined, there were no words or subjects that scared her, but about that, about sex, she never spoke.

"Five more weeks, I told myself, then four, then three, and at a certain point the truce would be over and one fine night we would find ourselves in that same bedroom, as before, with Corinne groping for me fearfully, saying, 'What do you say, should we do it?'"

TOMMASO LOOKED AT ME. "I'm embarrassing you."

"Not at all," I lied.

He poured more wine, brought the glass to his lips, but didn't drink; he held it poised for a moment, as if he were taking a breath to continue the story.

"One evening we invited Corinne's parents to dinner. 'They gave us this house,' she pointed out, 'not to mention everything else. And we've never officially invited them.' It made me smile, that qualification, 'officially,' typical of her family.

"She asked me at least ten times what I was planning to cook, she was worried. I surmised that she had greatly exaggerated my culinary skills in her father's eyes. I didn't reproach her. She had entered the last month of pregnancy and was plagued by leg cramps. She always seemed to be on the verge of collapsing.

"Her parents arrived with a bouquet of pink and white flowers. Her father handed me a bottle of red and ordered me to open it at once. I objected that dinner would be fish. 'I would prefer to drink this,' he said, 'if you would be so kind, Tommaso.'

"The food was fairly good, but Corinne persisted in winning me more compliments than necessary. At one point I met her father's gaze. He smiled, but it was a smile full of unspoken innuendos, as if to say: Look what we're willing to do for her, right?

"'You underestimate yourself, I'm always telling you that,' Corinne said when they had gone.

"'If he had been a little less enthusiastic, I might even have believed him.'

"She stared at me, wide-eyed and indignant. She got up heavily from the table and went into the bedroom.

"And then Ada arrived, two nights before the countdown had reached its end. We jumped in the car at four in the morning and less than an hour later she was in Corinne's arms, while I filled out forms on the floor below.

"She brought a joy that I hadn't expected, but it didn't last long, a few weeks, maybe a few months. I'm not saying that afterward I stopped being happy, it's not that. But the exhilaration over Ada's birth quickly faded, every day an ounce less wonder and an ounce more discontent.

"The truce with Corinne ended. We resumed being reciprocally offended, as if we had never moved from that evening with her parents. I was constantly asking myself whether I loved her or not, and if so, how much. You can go out of your mind wondering if you love someone or not."

TOMMASO PAUSED. He let the last allusion fill the air already saturated with revelations.

"In the evening I would hold Ada in my arms. I cradled her to put her to sleep and I could feel that she had imperceptibly gained weight. I looked at the color of her cheeks and thought I couldn't have generated her. So normal. So perfect. I examined every detail of her, studied her gray eyes, until what I was doing scared me. Then I'd put her back in her crib. If she cried, I let Corinne take care of it. And then I began to spy on her too. As if she were an enemy. Oh, Corinne got back at me later! She humiliated me every way she could. But still, it's nothing compared to the magnitude of my hostile thoughts during that year. There was the ungainly way she sat, there was the hair she didn't wash often enough and the crocodile yawns and the odd way she held her fork and her loud voice. There was only one way to stop those thoughts. By drinking just enough, my life in the apartment became tolerable again. At the beginning I would limit myself to having something before going home, at a bar along the way. I drained three glasses of rosé, as if it were medicine, then I got back in the car."

"It sounds like you're justifying yourself."

"Could be. Maybe you're right, maybe I am justifying myself. But I'm also telling you what happened to me, exactly as I told it to Bern one evening. He was very severe. He said it was shameful, that I didn't appreciate my good fortune. In fact, no, he didn't say 'shameful,' he used one of those adjectives of his that seemed chosen to hit home. 'Deplorable,' that's what he said. Then he added that I really didn't deserve a daughter if I couldn't be joyful over her. It had to do with . . . well, your issue. I already knew about it, but I assure you that before that comment of his I hadn't even connected the two situations."

"What issue?" I asked.

Tommaso remained silent long enough to make it clear that he wouldn't answer, his head slightly bowed.

"What issue?" I repeated.

"I shouldn't have mentioned it."

"You shouldn't have mentioned what?"

I felt like gripping those hands of his, so pale and limp, and crushing them.

"The insemination. Kiev. Et cetera."

I stood up. Medea abruptly raised her snout.

Tommaso turned to look at me, with no compassion and no remorse. Then he said, "Please sit down."

And since there was really no place else I could go, I obeyed him. Medea settled down too, her snout once again resting on her paws.

I said, "Apparently not all secrets are worth keeping."

"Bern and I told each other . . ."

"Everything, yes, I know."

Tommaso coughed, then he cleared his throat and went on.

"I kept a supply at home to get me through the weekend. Vodka mostly. In that respect I was different from my father: he didn't touch the hard stuff, just wine, which got him drunk more slowly and then left him completely wrecked. Mine was progress, in a way."

He gave me an ironic smile, but I remained impassive.

"I remembered Bern's words, so I invented a toast: To God's gifts and to deplorable men! I got so used to it that I still say it today. I repeat the phrase mentally. I don't know to what degree Corinne was aware of all this. Probably more than I was willing to admit. But she didn't say anything. Sometimes I caught her furtive, frightened expression again. That side of Corinne was a new development. If there's one thing I would not have guessed about her, it was that she could be intimidated. I said to myself: She's right to fear me.

"After the night I'd talked with him, I didn't see Bern for quite a while. At other times I would have worried, but for the first time I didn't care that much. Besides, all I had to do was adjust the dose of alcohol and it absorbed that disappointment as well.

"Until one day he showed up without notice. It was early summer. Corinne was at the shore with her parents and the baby.

"'I'll open a couple of beers,' I said.

"'I can't stay long.'

"'Is there someplace urgent you have to be?'

"Suddenly we were both struck by how foolish such distance was, the reciprocal wariness. I felt like hugging him; he noticed and smiled at me, then he sprawled on the sofa and said he would gladly accept a beer, provided it was icy cold. We sipped from the bottles for a while, not saying anything, as if getting acclimated to the intimacy. I felt good. At peace.

"'The mulberries have ripened,' he said at one point, and I visualized the huge tree at the masseria and us as little boys, trying to reach the fruit at the top. I was grateful to him for that image.

"'Will you do something with them?'

"But he ignored the mulberries. 'Teresa and I are getting married,' he said. 'In September. I wanted to ask you a favor for that day.' Now he'll ask me to be his best man, I thought, and I'll accept, of course I'll accept. I'll get up from here and give him a brotherly hug, as is appropriate for two adults in these circumstances.

"But instead Bern said, 'I'd like you to take care of arranging the refreshments. We don't have much money. We'll have to try to do it all cheaply, but you're good at these things.'

"'Of course,' I replied automatically, just as I had intended to do, though to a different question.

"'Teresa already has some ideas. Maybe it's best if you two meet to talk about it. I'm taking care of the rest with Danco.'

"When he left, the sun was sliced in half by the expanse of sea, a sweaty ball that radiated an orangey glow into the apartment. I remained standing until it was dark, then I acted with a resolve that made no sense. I turned on all the lights in the apartment, in every room, then I switched on every household appliance. Washing machine, dishwasher, air conditioners, vacuum cleaner, kitchen exhaust fan, even the blender at maximum speed. I grabbed a bottle of white wine from the fridge and left the door wide open so that the refrigerator too would start humming in pain. I went to sit on the sofa again, the bottle in my hands, surrounded by the vibrating noise of everything that had made my life newer and more respectable, of everything that had invaded and destroyed it.

"Oh, it was a special wedding all right! I arrived already well fortified. I had to make up an excuse to get Corinne to drive us from Taranto to Speziale. I had drunk something very sweet, some Baileys, I think; my stomach was churning. Luckily, Corinne was furious with Bern. 'To pick Danco as his best man instead of you!' she kept saying. 'It's typical of an asshole like him.' And she was even more indignant that he had asked me to work that evening. For once I was content to listen to her rail against the masseria, against all of you, to have her on my side. I put my hand on hers and kept it there even after she stopped talking.

"When you came back from the ceremony, happy and a little disheveled, just like the day I'd caught you coming out of the reed bed, the alcohol had not yet worn off. When the guests trickled over a few at a time to ask me where they should sit at the tables, I barely understood them. Once everything seemed under way, I allowed myself to leave my station and dance a little. We even danced together, you and I. Corinne grabbed me by the tie. I kissed her, a kiss more fervent than any other we'd exchanged. For a moment we stood unmoving among all the people. I remember thinking: All this can work, I didn't think so, but it can work. Starting tomorrow you'll change, starting tomorrow, sure. Bern was right when he reproached me. Then I left her there and went back to preside over the buffet.

"That's when Nicola came over. If it hadn't been for my momentary giddiness, he might not have caught me so off guard. And maybe I could have handled things differently. I found him looking for something under the table.

"'What can I get you?' I asked him.

"'Ah, there you are,' he said, straightening up. He looked a little spaced out. 'Where's the hard stuff?'

"I gave him the flask I had inside my jacket and he told me I was a bastard, that he knew he couldn't go wrong with me. He said it almost affectionately, then he emptied it in one swig and belched.

"'Here you go, waiter,' he added, still staring at me. 'Or, rather, no. It's not polite to tip waiters, is it? They're only doing their job.'

"The need to provoke me came to him just then, I'm sure of it. Something about me riled him. There were two bottles of wine open on the table, right between us. He appraised them, then knocked them over with his index finger, first one and then the other, like skittles. The wine streamed over the table, my pants, and my shoes.

"'Oops,' he said.

"'Shithead.'

"'So what, you have someone to buy you new clothes now, right?'

"I had no clue where he'd gotten that idea about my life, we hadn't been in touch for a long time. Only later would I realize that he had never stopped spying on us, all of us, when we were together at the masseria and then afterward.

"He said it was really a shame to see a waiter spill wine all over himself, and that it was a good thing Bern had chosen someone else as his best man instead of me. He knew me well, he knew exactly where to strike. I didn't even answer him, I just threw the napkin I was using to wipe myself at him, that's all, but he sprang at me like a wild beast. He grabbed one of the bottles and raised it in the air as if he were about to smash it over my head. He held it there for a few seconds, then he started laughing, as if it were all a joke.

"That's when Bern came over. He must have only witnessed the final scene, when Nicola laughed, because he didn't seem alarmed or upset. There we were, we three brothers, together, after all those years. Nicola wrapped an arm around Bern's neck. 'Here he is, the groom. Three cheers for the groom!' he shouted. 'Waiter, three glasses, right away. Let's drink a toast to the groom!'

"We actually toasted, with Bern seeming faraway and Nicola more and more revved up. Then he glanced around, as if looking for something. 'It's over there that we threw the stones, right? That very spot, I think. Yours, Tommi, went as far as that olive tree. Am I right? Is my memory correct, Bern?'

"'Nicola, not now,' I begged him.

"Bern still hadn't said anything.

"'Why not? Why not now? When do we ever get a chance to exchange a few good memories! So then, another toast to the groom! Refill the glasses, go on!'

"We drank again, slightly more strained.

"'So then, bridegroom, tell us.' Nicola held an imaginary microphone in front of Bern. 'How does it feel to promise fidelity in this cursed place?'

"Bern took a deep breath. He set the glass on the table and started to return to the dancing. But Nicola wasn't done. Suddenly he turned serious again and asked Bern: 'Does she at least know where it is she's getting married?'

"'We took an oath,' Bern said quietly.

"Nicola took a step toward him. 'Because if she doesn't know, I can always explain it to her.'

"At that point, it was Bern who stepped forward. He looked up at Nicola, without the slightest trace of fear or submission. He said clearly: 'If you say even one word to her, I'll kill you.'

"He said it coldly, and Nicola laughed. 'Let me remind you that I'm an officer of the law.'

"They faced one another for a few more seconds, framed by the ornate coils of the tiny lit bulbs. Then Bern again turned to leave. But Nicola still hadn't finished and hurled those words after him."

TOMMASO STOPPED TALKING. Maybe he was looking for a way to backtrack, to eat his last statement.

"Which words?"

"It doesn't matter now."

"Tell me what he said, Tommaso."

"He said, 'I heard that you and your wife are having some problems.' Bern didn't turn around. But he stopped, his arms slightly away from his hips. 'Maybe we were wrong that time. If you need some help, just give

253

a holler. We'll do like old times.' Even then Bern didn't turn around. Another few seconds went by, then he started walking again, very slowly, and disappeared among the guests. Later there was the cake and that speech by Cesare. All that idiotic nonsense about the Book of Enoch. Who could really understand him? Only we three: Bern, Nicola, and I. Because who else could the rebellious angels be, if not the three of us? Fallen from heaven, from that paradise that Cesare had created, and plunged into fornication. Damned for eternity. He took that opportunity to tell us that he had not forgotten, that he knew much more than we wanted to believe, and that as long as we persisted in keeping the secret there would be no hope of redemption for us. His Sermon on the Mount, his final lecture. It was a good party, sure. I had some cake and listened to Cesare and watched the fireworks explode. But I was no longer able to enjoy any of it. And my good intentions for the following day had already dissolved."

"What did you mean that Nicola had been spying on us the whole time?" I asked. I was still stuck at that point. I had heard the rest, but without really grasping it.

"He kept watch on us while we occupied the masseria with Danco. And even after that, I guess, when you and Bern remained there alone."

"And later still, as well," I said, now talking more to myself than to Tommaso.

After Bern left, having burned our entire store of wood in a single night, and I was left to sleep alone in the house, surrounded by the sounds of the countryside and by a silence even more frightening, I often had the impression that someone was out there, that someone was watching me. There was no need to go out looking for him to be certain, or even to listen for the sounds any more intently than I already was. But I thought it was Bern. His pride wounded, yet still faithful to the commandment that Cesare had given us on our wedding day.

I must have spoken some of those thoughts aloud, because Tommaso said: "No, it wasn't him. Bern went back there only once, as far as I know. When he was already living here. And he found Nicola's car parked on the side of the road. Nicola wasn't around. That's how he confirmed that you two . . ."

"That we?"

"It's none of my business," he said shortly. "Anyway, I too tried to convince him that it wasn't like that."

In the other room, Ada's breathing had changed. It was heavier now, similar to that of an adult.

"I'm not following any of this anymore," I said.

"If you would just let me go in order," Tommaso said, his voice firmer. His right hand went to his mouth and he tapped his bloodless lips repeatedly, as if trying to coax out the next few sentences.

"Remember when we found the solar panels trashed? We thought it was some farmer's spite work, some rival, who knows. But it was Nicola. Him together with some of his colleagues."

"You're just saying that because you hated him. Like Bern."

Tommaso shook his head.

"And how would you have found out?"

"Nicola told me himself. A few weeks after the wedding, I saw him again at the Relais. He appeared just like that, out of the blue. I went over to a table to take the orders and there he was, smiling, wearing a light brown sports jacket. He introduced me to the three men who were with him, as if I were the reason why they had come from Bari. It was nearly winter, because the tables were set inside. November, maybe? It doesn't matter. He pulled me to him by the arm and said to his colleagues, 'This is my brother.' Then he explained that we had neither a father nor mother in common, no relations, but that was irrelevant, because we were even more intimate than two blood brothers. He said, 'We jacked off together,' and his friends thought it was very funny. One of them quipped that I must still devote quite a bit of time to it, since judging by my coloring I didn't leave the house much. Then they laughed even harder, Nicola included. But when they were finished he pointed a finger at the one who'd made the wisecrack and said that no one should make fun of his brother. I didn't get it. Last time, at the wedding, he'd done nothing but needle me, and he and Bern had nearly come to blows; now here he was in the Relais's dining room playing the part of big brother in front of the other policemen in civvies.

"They ordered two bottles of Veuve Clicquot. Nobody ordered champagne at the Relais, the price was off the charts. I worked all evening feeling slightly uneasy. I had the impression that Nicola never took his eyes off me. But maybe it was the other way around, maybe I was the one who couldn't forget his presence. I watched him from a distance and tried to reconcile the cheerful man sitting at the table with the one who'd been raging out of control at the wedding, and again with Nicola the boy.

"The room emptied and they were the only ones left. It was very late. I'd left them two bottles of grappa and they seemed intent on finishing them. Nacci motioned me to join him. 'Your friends want to play cards. Did you tell them about it?' Evidently Nicola remembered what I had told him during our time at the Scalo. Nacci glanced at them for a moment.

"'I don't know what you were thinking. They're cops, for God's sake! Anyway, it's fine with me. Have them move into the card room.'

"'I should go home,' I said.

"He stepped closer. 'Listen to me closely. You got us into this situation. Now your friends want to play cards, after all the champagne they drank. And we won't disappoint them, will we?'

"So I ended up being the dealer for Nicola and his buddies. Blackjack, until five in the morning. They lost at least two hundred euros each, but when they left they were euphoric. I walked them to the car. A ground fog had risen from the fields. Nicola grabbed my head and pressed a chaste kiss on my lips. He also said something affectionate to me, downright soppy. He was really quite drunk at that point.

"After that night they started coming every Saturday. They had dinner and then played cards. Nacci started treating them like guests of honor, and often lingered with them. He paid me for the extra hours with a percentage of what the bank won, as he used to.

"Corinne did not look favorably upon those nights. She knew about the cards, but I didn't tell her about Nicola. It was as if the most unacceptable part wasn't the gambling or the drinking, or even staying up so late and then sleeping for most of the following day, the only day of the week that I could have spent entirely with her and the child; no, the most unacceptable part was that those evenings centered on one of my brothers.

"After a few weeks she couldn't take it anymore and decided to confront me. It was Sunday afternoon and I was still in bed. Corinne came into the room but kept her distance.

"'Why do you need to do it?' she said.

"'It's extra money. It comes in handy.'

"'We don't need more money. We have more than we spend.'

"'No. You have more than we spend. My account always shows the same amount.'

"I was inexcusably icy when I said it. I did it on purpose. She stood in the middle of the room, while I lay in bed as if it wasn't even worth the effort to pull myself together. The light pressed against the drawn curtains, spilling out at the edges. I think Corinne started crying, in the dimness I couldn't be sure; in any case I remained lying down, until she left."

TOMMASO WIGGLED a foot under the blanket, causing Medea to twitch, though she did not wake up. He smiled faintly at her.

"They knew how to party. Nicola and his pals. One night I caught two of them in the bathroom taking turns snorting a line of coke. They nodded their heads for me to join them but my response was to go look for Nacci. I told him what I had just seen. I think part of me still wanted to get rid of them.

"'So now you're being a moralist?' Nacci replied. 'Let them have a good time. Or do you want to report the police to the police?' He went off chuckling at his joke. His reaction had the effect of a perverse blessing on me. From that night on I let myself go. I would often play poker, with my own money, I drank if there was alcohol, and I joined the pilgrimages to Nacci's private bathroom. It was in that bathroom that Nicola told me. Not because he was sorry, or to provoke me. There was something fiercely sincere between us at that point, as if all outstanding accounts had been canceled and our brotherhood, always obstructed by Bern, had found the chance to express itself.

"'Remember the solar panels? It was me and Fabrizio. It took us almost two hours.'

"'Why?'

"'You hadn't called me, not even once. In all that time, not once. I'd see you. I saw what you were doing, when you were all together in the evening under the pergola. That place belonged to me too.'

"A few days before Christmas they rented the Relais, all of it, a party in grand style. I helped Nicola with the preparations; by now I was becoming an expert in planning other people's parties. We decided on a fish-based menu, we found a deejay, and one morning I accompanied Nicola to a wholesale store outside Gallipoli, where we bought up a supply of alcohol and a quantity of party gizmos: glow sticks that became fluorescent when broken, headbands with teddy-bear ears, silver and gold masks with elastic bands, and firecrackers. We showed up at the cashier's desk wearing the masks, like a couple of kids. I was happy.

"On the way back, Nicola told me about the girl he was seeing. Stella.

He elaborated on certain very private details, maybe to impress me. He said they had this agreement between them: one month apiece, each of them had absolute power over the other. When Nicola was in charge, he could order Stella to do anything at any time, and vice versa. Obviously almost all the demands had to do with sex. Often they involved other couples, or unattached men and women, for a fee. He was neither arrogant nor funny as he talked about it. In his mind, it was something very serious.

"'Do you like those games?' I asked him at one point.

"Nicola narrowed his eyes to stare more intensely at the road that wound through the vineyards. He said, 'If I don't do that, I don't feel anything. Nothing at all.' He spoke those words with great sadness. Then he added, 'Isn't it like that for you too?'

"But I avoided the question. 'Have you introduced her to Cesare and Floriana?'

"He burst out laughing. 'Have I introduced her to them? Hell no. No way!'

"'And do you still think about her a lot?'

"I still couldn't get over the fact Nicola and I had ventured into such intimate territory. But for years I'd misjudged him. When I asked him if he still thought about her a lot, I was referring to Violalibera. Instead Nicola replied: 'She's married to him now. What can I do?'"

I STOOD UP ABRUPTLY. "Do you mind if I open the window for a moment? It's stuffy in here."

"Whatever you like," Tommaso said.

The cold air hit me in the face; it smelled vaguely of the sea, although you couldn't see it from there, only other buildings, all dark. I breathed in that fresh air for a few seconds, then closed the window and went to sit down again. Tommaso waited patiently, somewhat lost in thought.

"Do you feel okay?"

"Yeah."

"I can stop if you want."

"No, go on."

"You should drink some wine too."

"Go on, I said."

"There were about eighty people at the party, all policemen with their girlfriends. After dinner, when the deejay turned up the volume, they all got up to dance. Nacci stood in the doorway counting the buckets of Veuve Clicquot that went by him.

"Nicola and his little group climbed up on a table and danced on that improvised stage. Stella was there too. Seeing her there, you wouldn't have said she was capable of what Nicola had described to me.

"I took advantage of the coke supplies available in the bathroom and downed countless glasses left half full, before putting them in the dishwasher. I remember thinking: Corinne's father should see me now; I wouldn't be able to touch the tip of my nose with my eyes closed, but I can still carry a tray with thirty glasses. I found myself standing on the table, a whole new perspective on that dining room that I had walked through thousands of times.

"Nicola was dancing behind me. I ended up crushed between him and Stella. Then other people climbed up. I was moved by the packed contact with all those bodies.

"After that I have a kind of memory gap of a few hours. I remember going into a long, narrow apartment, like a corridor, where there was a wall painted black that you could write on with colored chalk, and I wrote something on it that the others found funny. Outside it was already lightish. There were five of us, the next morning, at least.

"I woke up on the carpet and went down to the street. It seemed a Sunday morning like so many others, bright and warm for December. I realized I was a few blocks from home. I went into a bar and in the bathroom I tried to pull myself together; my vision was slightly clouded.

"When Corinne saw me, she couldn't speak for a few minutes. She kept pacing nonstop through the rooms.

"'It's eleven o'clock,' she said finally, as if she had counted the hours that had gone by, one by one.

"'The party ended late. I slept at the Relais not to wake you up.'

"'Not to wake me up? Really? I called the Relais at eight. They told me you'd left a long time ago.'

"I went over to her. I touched her arms, but she stiffened as if I'd scared her to death.

"She said, 'I have to go out. And you have to change Ada. You have to watch her.'

"Then she took her things and left, still in a kind of trance.

"I was in a state of confusion. And tired. My hands were shaking. I was well aware of the effects of a hangover, and if it were just that . . . But there was all the cocaine as well, I had no idea how much even. And those memory flashbacks of the night before. I sat down on the couch and probably collapsed immediately. What woke me was Ada crying in her little room. By then she was really shrieking, I don't know how long it had been going on. I lifted her out of the crib and held her in my arms. I was starving, I hadn't eaten anything since before the party, but when I tried to put her down on the floor, she instantly started screaming again, so I picked her back up. I put a pot of water on to boil, looked in the refrigerator for some leftover sauce, found the pasta. I held Ada with my left arm, I'd done it hundreds of times. Maybe she made an abrupt move. She lurched backward. I'd left the cabinet door open.

"There was so much blood, I couldn't even see the wound. The emergency room doctors said she'd been deprived of oxygen for several seconds, not because of the blow, but because she'd screamed so hard.

She'd suffocated from fright. Corinne had already arrived and her parents had come, and some other people—I didn't even know why they were there. Someone brought me tea from the vending machine, which tasted of lemon concentrate; I took a sip, then left it to cool. I kept wondering why Corinne wasn't pouring out her anger at me. The doctor spoke with her, then left without adding anything about being confident, about being hopeful. Cesare came to mind and I felt a wrenching nostalgia. At a time like that he would have said the words that were needed.

"Toward evening the swelling of Ada's head had gone down. Corinne went home to get some rest. The nurses asked me to step out of the room for a moment. Corinne's father was in the hallway. He put a hand on my shoulder. He spoke kindly, a true diplomat, I thought. He gave a brief recap of the events, as if I could have forgotten them. He said that he had never seen his daughter as unhappy as she'd been recently, not even during the worst years when she was a girl. He never called her by name, he always said 'my daughter.' It was time for me to undergo treatment, my problem had become a serious concern. My problem. 'Right now you're sorry, you're certain you want to set things right and that the recent scare will give you the strength to do it. But that's not so. You could go back to her and promise her that from now on it will be different, but you and I know that's not true.'

"Then he explained the solution that he had arranged in the last few hours, or that more likely had been ready for some time and had only been waiting for the right opportunity. He explained about the apartment that had become vacant, this apartment where we are now. Not to let it slip away, he had already paid a few months' rent, and he wouldn't ask me to repay him. I could consider it assistance to start over again. Naturally, I would continue to see Ada, everything would be decided very civilly before a judge. I might have to put up with his wife being present with us, at least at the beginning, as long as it took for me to get back on track. However, if they had wanted to prevent me, it would have been all too easy, given how things had gone, right? But you don't punish a man for an accident. You don't expunge a father just because he has weaknesses. Who doesn't have any?

"In exchange for clemency, all he asked was that I do him the favor of not reporting any of that conversation to Corinne, that I take full responsibility for the plan. She would suffer a little at first, but ultimately she would appreciate it. Because women can recognize when men have the courage to act on their decisions, he said. If he were me, he would wait a couple of weeks, allow enough time to recover from the scare. If he were me, he would let New Year's go by, but no longer, because later it would be more complicated for everyone. If he were me . . . So I let him be me."

TOMMASO PAUSED AGAIN. He seemed to be pondering something, and finally he asked me to bring him a cigarette.

"Won't it make you feel worse?"

"No. It won't make me feel worse."

I went to the other room, found the packet, and returned to the bedroom. I lit Tommaso's cigarette, then mine. We used his glass as an ashtray.

"It was what I wanted, deep down. To get out of there. To be rid of Corinne and her expectations. My debt had been repaid for some time. Yet the first few weeks were the worst. When I wasn't at the Relais, I was at the bar across from the harbor. I'd meet Ada there, together with Corinne's mother.

"'Why don't we go to your house?' the woman urged me after a few times. 'It's important that the child see where you live now. So she won't think her father doesn't have a home.'

"'Her father doesn't have a home,' I replied, and she did not insist.

"Those meetings were torture. Ada may have been the only one who wasn't aware of it. She wandered among the tables in the bar, and the customers smiled at her. Corinne's mother always brought toys, toys that I myself had bought before I left. But she didn't know it. Who knows what Corinne had told her. She explained how to play with them, but I preferred to watch. As soon as they left, I would order a drink. It went on like that for a couple of months, though, looking back, it seems like a very long time. Sitting in that bar, watching the virtual cards scroll by on the slot-machine screens. Then Bern showed up. Out of nowhere, like always, there he was in the bar. The place on earth least suited to him. He scrutinized the surroundings for some time before coming over.

"'Let's get out of here,' he said.

"'Why?'

"'Let's just go. Where's your house? I have my bag in the car, but I have to take the car back to Danco before evening.'

"And so he kept the promise he'd made to me all those years ago at the masseria, the night we'd stood in front of the window and he'd promised to take care of me.

"The next day we started stripping off the stained wallpaper. We took the most battered furniture to the dump and bought new pieces at a warehouse. Bern talked a lot, practically nonstop. In recent days he'd been living with Danco in a kind of encampment. 'The Presidio,' he called it. Since the Xylella pandemic had begun, Danco and some others had mobilized to prevent infected olive trees from being cut down. They had formed some sort of militant group and slept in camp tents around a farmer's cabin. It was the farmer who'd convinced them that cutting down the trees was useless, that there were undoubtedly some profits to be made behind the operation. He was treating the infected olive trees with copper sulfate and lime. Bern was animated as he told me all this. The voice was his, but it was Danco talking. And meanwhile, he tore off strips of wallpaper and painted the bare walls a ridiculous pink, which that little girl would like." He stopped. "Why are you looking at me like that now?"

"I'M NOT LOOKING at you in any way."

Tommaso crushed his cigarette out in the bottom of the glass, then held the makeshift ashtray in his lap.

"Yes, you are. You're looking at me because I haven't said anything about Bern and you. I haven't said anything about what he told me. But he didn't talk about it much. That's the truth. Only one evening, while we were eating Chinese food, sitting on the floor, he said: 'Chasing after an egoistic desire tore us apart.' Then he blamed your doctor. He had gone to see him a few days earlier. I think he must have made a scene, that he threatened him in some bizarre way. Saying he'd tell everyone what he was up to, talk to the newspapers about it."

"Did Bern tell you that? That he went to Sanfelice and threatened him?"

"He was a little ashamed of it, I think. Or maybe not. Anyway, he must have been distraught when he did it, so he didn't elaborate much on the details. All he said was that he'd burst into the clinic right in the middle of a visit, that the secretary had tried to stop him, and that he had let the doctor have it. We were sitting on the floor, all smeared with pink paint, passing the container back and forth, the Chinese noodles all stuck together in it. Then Bern said, 'Teresa slept with him. With Nicola. I saw his car parked outside the masseria. A few nights ago.'"

"And what did you say?" I asked.

Tommaso turned toward the window.

"Didn't you say anything to him? You had already talked to Nicola at that point. You knew that he came to the masseria to spy. Why didn't you say anything?"

Tommaso lay still, as if by not moving, my voice would pass over him without touching him. I grabbed him by the arm, but he pulled away brusquely.

"Look at me, Tommaso!"

His eyes had changed, they were open wider now, full of rage, or terror.

"Why didn't you tell him the truth?"

"I couldn't be sure," he replied in a faint voice.

I took a deep breath before hurling my accusation: "No. You didn't tell him what you knew about Nicola because you wanted to keep him there with you. You kept quiet and let him go on believing what he already thought."

Tommaso's eyes were still wide open, his gaze riveted on me.

"Isn't that right?"

"I guess so."

I stood up, went to the kitchen, and got two clean glasses. I poured some wine for both of us. I wanted to put on my coat and leave, not listen to anything more. But not this time. I would hear him out, to the very end. I went back to the room and gave Tommaso the wine. He sipped it slowly.

"And then?"

"Nothing. At least not for a while. In a couple of weeks the apartment was ready to welcome Ada. Corinne's mother came to check. She stood aside and watched as Bern, 'Uncle Bern,' twirled Ada around. Bern doted on Ada, and she on him. Anyone else would have made me jealous, but not him. They were happy months. The best, maybe."

"It sounds like a dream come true," I said spitefully.

Just then Tommaso started crying. Trapped in the bed as he was, he covered his eyes with one hand, sobbing. I watched him for a while.

"I'm sorry. I shouldn't have said that."

He wept almost soundlessly. I waited for him to take his hand away from his face.

Then he took a sip of wine. Wiped his mouth with the back of his hand.

"Bern took me to the encampment. The infected trees were marked with a red cross painted halfway up the trunk, waiting to be chopped down. Danco and his group swore they wouldn't let anyone get near them.

"In the evening we cooked hamburgers outdoors on a greasy black grill. There wasn't much to do, to tell the truth. No immediate threat to counteract, no plan. Many of the young people were university students and they lay sprawled out with open books on their stomachs. When it got dark, they lit a bonfire. Danco gave one of his speeches. A somewhat disjointed speech, in fact. But they were all younger than him, attracted by his citations. I wanted to go home, but Bern insisted on sleeping there. He went to Danco and Giuliana's tent. I found myself with two guys, inside a sleeping bag that smelled like sweat.

"The next day, Bern and I left very early, everyone was still asleep. We drank some cold leftover coffee from a thermos.

"'Did you like it?' he asked me in the car.

"'It's a shame about those olive trees.'

"'It's not a shame. It's a crime,' he said, keeping his eyes on the road.

"So now I spent some nights at home with Bern and the child. I spent some nights at the encampment. And I spent those wild nights with Nicola and his buddies. Separate lives, which had nothing to do with one another: my specialty.

"I was at the encampment when a couple of reporters came to interview Danco. The Xylella had spread farther north. The protocol called for cutting down olive trees within a hundred yards of every infected specimen, but this would have meant deforesting the entire region. Danco blew his top, shouting at the journalists that it was all a pack of lies; he spoke about multinationals and lobbies. To all of us it seemed effective.

"That evening we gathered in the farmer's house. The clip was one of the last items on the news. Danco's interview had been cut to a few seconds, in which he claimed that Xylella was a media invention. In the video he appeared outraged and maniacal. After him a ministerial official was questioned, who provided precise data on the extent of the disaster.

"We returned to the tents with a sense of defeat and frustration. Bern went to sit under one of the olive trees. He remained there, staring wide-eyed, until late at night.

"In June, Ada turned three. She celebrated it with Corinne and her grandparents, then with me and Bern. We made a cake and got all dressed up. We acted a bit ridiculous. At the end of dinner, I turned off the lights and brought in the cake. We sang shamelessly and Ada was radiant. Bern had bought her wooden blocks carved with letters and numbers on them. Ada didn't pay much attention to them, and he was disappointed. He darkened even more when he saw her so excited by my gift, a doll. 'It's all plastic,' he said angrily.

"Then he stormed out of the apartment, leaving us there. He returned a few days later and we didn't say a word about the birthday. It went on like that all summer long and throughout the fall. Bern was spending more and more time at the encampment, but every so often he came over. He no longer talked to me about what was happening over there and I wasn't very interested. Even when he showed up with his shoulder bandaged, he was vague, but that time he stayed longer. In retrospect it all seems so obvious, I should have realized what was coming.

"In December the bacterium reached the Relais dei Saraceni. Actually, Nacci never had the trees analyzed, he merely studied them and showed me the yellowed branches. They could have been that way because of the sun, the drought, but he had decided that part of the olive-tree reserve had to be cleared out. He had already made arrangements.

"'The special regulations on Xylella allow him to,' I told Bern and Danco one evening.

"'But why?' Danco said, worked up. 'It makes no sense, it's a loss for him too.' He made some mental calculations that didn't add up, and he couldn't see what Nacci got out of it. So I added, 'He's cutting down the olive trees because he wants to put in a golf course.'

"Silence fell. Danco and Bern looked at each other. This was the action they'd been waiting for. A grand, momentous, concrete move. They'd had enough of removing the red crosses painted on the tree trunks and drinking second-rate beer with an illiterate farmer, waiting for who knows what.

"After that evening Bern didn't show up again. A couple of months went by without hearing from him. A couple of months, yes, because when I found him at my house, by surprise as always, it was already February. I immediately noticed the big box next to the sofa, and asked him what was in it.

"'Some stuff,' he replied evasively. 'Don't touch it, please, I'll get it out of here right away.'

"Naturally I looked inside as soon as I was alone. I peeled off the scotch tape slowly, so I could put it back exactly as it had been. There were bags of ammonium nitrate, which I was familiar with because we used it as a fertilizer at the Relais.

"I was at home with Ada when Bern came back a couple of weeks later. He didn't even take off his jacket when he came in but went straight to the carton. The next day the bulldozers would be at the Relais.

"'Will you be with us?' he asked.

"'You know I can't. I have a job there.'

"At that moment I realized that I should have thrown everything away while he was gone.

"'Leave it here,' I said.

"'Will you be one of us or not?'

"'Leave it, Bern. It's a stupid idea.'

"He lowered his head. 'From now on it no longer concerns you, Tommi.'

"I sat on the box, like a child.

"'Get off,' Bern said.

"His voice changed. The severe tone had been replaced with the same emotional, regretful voice with which he'd begged me not to read the Gospel of Matthew under the holly oak, the same one with which he'd asked me to steal from the cash drawer at the Relais.

"He took my hands and made me get up. Then he bent over the box. 'You can come with me, we can be together this time too. It's our most important mission.'

"But it wasn't like that. That wasn't my most important mission. Ada was sitting on the couch, mesmerized by the cartoons.

"'No,' I said.

"Bern nodded, the box poised in his arms, the door already open.

"'Call the elevator, would you?'

"I stepped beside him. I pressed the button. In the time it took for the elevator to arrive, we didn't say another word. The doors opened, Bern got in, the doors closed behind him. After that, I never saw him again."

TOMMASO SUDDENLY pushed the sheet aside, uncovering his pale legs. He stood up.

"Be careful," I said.

He seemed to have suddenly gotten a grip on himself. He padded out of the room barefoot and went into the bathroom. I heard the stream in the toilet bowl, then flushing and water running, for a long time. There was no need for him to add anything else. I knew the rest from his deposition at the proceedings against Bern and Danco. I knew it from the testimony of all the witnesses and from the newspapers' reconstructions.

That night Tommaso had called Nicola. He had panicked and didn't know who else to call. Maybe Nicola would be able to talk some sense into Bern and the others without resorting to arrests. Like a friend. Like the brother he was.

Nicola went to the Relais with his colleague, Fabrizio, both of them off duty, both of them armed. The bulldozers were already there, ready to move into action, and the guys from the encampment had formed a human cordon, holding hands, with caps lowered over their eyes and scarves in front of their mouths against the cold.

Nicola and Fabrizio arrived just when Nacci was putting his hands on Danco, him first of all, of course, reaching for his face to pull the scarf down from his mouth. Danco reacted by shoving him and Nicola separated them. He said he was from the police and grabbed Danco's arms to handcuff him. Then Bern leaped at his brother to free his friend, and Nicola's colleague, Fabrizio, lunged at him. Nacci ran back toward the Relais.

The chain of activists had broken at several points. The two men in the vehicles started up and proceeded through the breach that had opened. One guy panicked—although during the trial it was never clear who he was—and set off one of the bombs. The explosion wasn't very powerful, but it sent the activists scattering among the olive trees.

Nicola and his colleague pulled out the guns. Fabrizio ran after the group and Nicola found himself facing Bern and Danco, the three of them alone.

Only the man on the bulldozer saw any of what happened later, confusedly, through a pall of dust and smoke that had not yet dissipated.

He saw Bern on the ground and Nicola kneeling on top of him, with his gun aimed. Then he heard a thud, not a gunshot, a muffled clunk. Nicola lay stretched out and Danco stood beside him, still holding the spade in his hand; he gripped it tightly for a few seconds before tossing it away.

At that point the man got out of the bulldozer to help Nicola. By the time he reached him, Danco was already on the run, while Bern stared down at his brother's body, incredulous, stunned. The man tried to hold on to him at least, but Bern also started running, down through the sloping grove of olive trees that would soon no longer exist, transformed into a golf course with spongy bright grass beneath the sky.

TOMMASO CAME OUT of the bathroom, but lingered in the living room for a few moments. Watching Ada sleep, I thought. When he returned to the room he smelled vaguely of toothpaste.

"We can sleep a little," he said.

"I'll go now."

"It's too late to go. Stay here. That side of the bed isn't as filthy as you think. Get down, Medea, move."

I was tired. If I got in the car, I would have to struggle to keep my eyes open all the way home. And maybe I didn't feel like waking up in a few hours, Christmas morning, alone again, after all I had heard.

Meanwhile, Tommaso was on his hands and knees on the mattress, completing the job of brushing Medea's hair off the sheets.

"That's it, done," he said. "And it's been at least a week since I've seen a bedbug."

"What?"

"I'm kidding. Relax."

He grabbed the pillow that he'd kept crushed beneath the one under his head most of the night. He tried to plump it back up to a respectable shape, to no avail.

"It's fine like that," I told him. "Don't worry about it."

He lay down on his side of the bed, very close to the edge to leave me as much room as he could. I took off my shoes, keeping my shirt and jeans on, and slid under the covers.

Tommaso had his back turned to me. He was still, as if he were already asleep, but he wasn't sleeping. My rival from the beginning. I put a hand on his shoulder; I had no right to do it and it wasn't something I would ever have thought I'd do with him, but I did it. He left it there for a few moments, then he put his own hand over it. Then we managed to sleep a little, just a few hours, but a deep sleep, as I hadn't had in years. The lamp was still on next to me. Outside, dawn broke, but I didn't see it.

PART THREE

LOFTHELLIR

6.

What I remember most about the morning the police arrived is the silence. A silence different from usual, as if even the birds were struck dumb and the lizards frozen in the grass at hearing those words that changed everything: *It appears your husband is involved in a murder . . . A police officer. His name was Nicola Belpanno.*

The policeman asked my permission to enter the house. I didn't see any reason to prevent him, yet I didn't immediately move to let him pass, he had to ask a second time and then slip sideways through the space between my shoulder and the doorjamb. His partner came in after him, embarrassed, his head lowered.

I looked at the room as it must have appeared to them: the table in disarray from the night before, set for one, my muddy boots tossed on the carpet, the blanket crumpled up on the sofa. All signs of neglect on the part of someone who never expects visitors.

"Can we go up?"

"I haven't made the bed yet," I replied stupidly.

I leaned against the fireplace wall. I wanted to say that there were no secrets up there, nothing they were looking for: Bern hadn't been in those rooms for a long time, even though I had imagined him practically every night before falling asleep alone. I'd pictured him crossing the space with his long strides, and I'd talked to him, talked to him out loud. But I stood there watching the policemen move silently about and then take the stairs.

It appears your husband . . . His name was Nicola Belpanno.

I could almost understand the sentences separately, but the connection continued to elude me.

I did not offer the cops a cup of coffee or a glass of water. It didn't occur to me.

When we were at the door again, the only one of the two who seemed authorized to speak said: "I believe we'll be back for a more thorough inspection. Maybe even today. I'd be grateful if you would avoid going anywhere in the next few hours."

Then they left.

I sat in the swing-chair. Its shaky frame creaked, even though I didn't feel I was moving. A strange form of shock, entirely new, was taking hold of me.

. . . involved in a murder.

Around nine o'clock the telephone in the house began to ring. Here we go, I told myself, now it's starting.

It was a girl from a sales call center. I let her finish, registering every-thing she said, every unessential bit of information about a discount on sports channels and a leased decoder. Then I told her that I didn't own a TV and something in my voice must have frightened her because she cut it short.

I stared at the silent phone for a while, as if waiting for the right call. Then I went back to sit under the pergola. The policeman had advised me not to leave, and I wouldn't. I would stay right there, until the ridiculous version of events I'd heard at dawn proved to be a hoax.

His name was Nicola Belpanno.

THEY RETURNED in the early afternoon, three cars, tires squealing unnecessarily before they parked. They had a search warrant and a different attitude than they'd had that morning. Now they were more determined, aggressive.

I went to sit under the holly oak. I stayed outside while each object was touched, turned over, opened, emptied. From the bench I noticed that some of the leaves were dotted with yellow. I picked one off and examined it against the light.

The same officer with whom I'd spoken in the morning joined me and sat down next to me. "Let's start again from the beginning, all right?"

"Whatever you say."

"This morning you stated that your husband hadn't been in this house for a long time."

"Three hundred and ninety-five days."

He looked surprised. Obviously he was surprised. On the night of our wedding, I thought, Bern was standing where he was now sitting.

"Should I conclude from this that you and your husband are no longer together?"

"I suppose you could conclude that."

"Yet at the registry office he is still listed as residing here. You did not initiate separation proceedings."

At that point I should have explained to him that the separation from Bern had in fact been announced. Announced by his burning a woodpile in the middle of the night. If he looked closely, he would still make out the scorched spot on the ground. And I should have explained to him that Bern could not change his residence for any other place in the world, because his soul was lodged here, among these trees, among these stones. Instead I was silent. The policeman tapped his pen on his notebook.

"Can you tell me where your husband was living during the past year?"

I lied, just as I had done a few hours earlier, in response to the same question. But whereas in the morning I had done it out of a vague precautionary instinct, I now lied deliberately to protect Bern. Whatever he had done.

"I don't know."

From that moment on, the grilling became more pressing. The agent had made an effort to be friendly, but it was clear that we were not on the same side. Was I aware of my husband's involvement with extremist environmental fringe groups? Did I also have associations with those elements? Were there places my husband habitually frequented, places he spoke of often? People he may have named? Had I ever seen him manufacture weapons? Was he interested in building explosive devices before?

No, no, no, my only answer was no. Seen from a distance, the policeman and I must not have seemed so different from the boys who sat beside Cesare in turn, with him doing the talking and me being silent, staring straight ahead or at my feet, a few monosyllables dragged out of me from time to time.

"Mrs. Corianò, I advise you to cooperate. It's in your own best interest."

"I am cooperating."

"So Bernardo Corianò is not linked to any extremist groups."

"No."

"And Danco Viglione? What can you tell me about him?"

"Danco is a pacifist."

"You talk about him as if you knew him well."

"We lived together. Here, for two years."

"I see. You, Corianò, Danco Viglione, and who else?"

"Danco's girlfriend. And another couple."

"Giuliana Mancini, Tommaso Foglia, and Corinne Argentieri."

"If you already knew, why did you ask me?"

But the agent ignored the question.

"You see, I find it very strange that you describe Viglione that way. Calling him a pacifist, when he's actually a convicted offender."

I felt it hard to breathe. "A convicted offender?"

"Ah, you didn't know that?"

The policeman flipped back a few pages through his notebook. He read: "For aggravated assault in 2001. Resisting an officer in 2002, in Rome. He and others stripped naked during an international summit. Odd, isn't it? Your housemate spent a few nights in custody. You weren't aware of it, I guess."

Someone was rummaging around in my bedroom, I saw him pass from one side of the window to the other. All he would find was loss.

"As for Giuliana Mancini," he went on, "the lady was arrested a couple of times with Viglione, but has also been charged with computer fraud. At the moment she too appears to be untraceable."

He straightened his shoulders. He set the notebook facedown on his lap, as if he were laying down a weapon.

"I'm curious. What exactly did you all do here together?"

"We harvested the olives. We sold our products at the market."

We were realizing a utopia. But I didn't say that.

"You were farmers, in short. And your husband, Corianò, is he a pacifist too?"

"Bern has his convictions."

"Explain that better. What exactly does he believe in?"

What indeed. He had believed in everything and stopped believing in everything. At that point I no longer knew.

I said, "He has a lot of faith in Danco."

The policeman looked at me, a flicker of triumph in his eyes. If Bern was a follower of Danco and Danco was a previous offender, then Bern too must be a dangerous individual. Answering him like that had been a mistake, but it was too late now. The agent was silent, maybe waiting for me to reveal something more, to go further, but I didn't say another word. Under the holly oak the air smelled like resin.

"How did he die?" I asked him finally.

"They bashed his skull. With a spade."

He used that brutal expression on purpose, I think, to get even with me for being reticent. It worked, because the image planted itself in my eyes: Nicola's head bashed in by a spade. It would never go away.

"Have you already talked to his father?"

"With Belpanno's father? Someone is with his parents right now. Why do you ask?"

I looked him in the eye.

"Do you know him?" he asked.

He looked off balance, as if he realized he'd been talking to the wrong person the whole time.

"Nicola and Bern are practically brothers. They grew up together. You people think Bern hurt Nicola, but you're wrong. His father, Cesare, will confirm it."

The agent told me not to move. From the holly oak I watched him walk away, then talk on the phone, plugging up his free ear with a finger. He did not come back to ask me any more questions.

After that they left. The same deafening stillness of the morning. I opened the goat's pen and watched her step out and idly graze the winter grass. She was looking for the harebells hidden among the blades.

I went into the house, imagining that I'd find it topsy-turvy, but it was orderly, a somewhat chilly tidiness that wasn't me, as if by putting things in place the cops had wanted to reproach me for my neglect. I sat down at the computer. The news appeared as the lead story on the *Corriere del Mezzogiorno*'s website: "Policeman Fatally Wounded During a Tree-Felling Demonstration. Suspects on the Run."

You could click on the title story or on one of the in-depth analyses: *The location of the conflict—Interactive map of Xylella—A life in government service.*

No mention of Nicola and Bern being related. I started reading the main article, but by then I was so shaken that I had to get up, go outside, and pace back and forth for several minutes.

When the phone rang again I ran to answer it. It was strange to hear my mother's voice. Since Bern was no longer living here, since the obstacle of Bern had been removed, we talked at least twice a week, but that wasn't the usual day, that wasn't the scheduled time.

"Oh, dammit, Teresa! Dammit!"

She was crying. I asked her to please stop. A very fragile balance was at stake. Something massive and irreparable was ready to explode in me and I knew it would happen if I kept on hearing her sob.

"They're talking about it on the radio too," she said.

"Of course," I replied, but meanwhile I was thinking that my parents never listened to the radio. Maybe things had changed since I was gone. Maybe now they did.

"Come back here, Teresa. Come home. I'll go to the agency and get you a ticket."

"I can't leave the area. The police advised me to stay nearby."

Mentioning the police triggered a fit of hysteria. But this time it didn't produce any reaction in me.

"Isn't Dad there?"

"He went to bed. I persuaded him to take a Xanax. He was beside himself."

"Mom, I have to go."

"No, wait! Your father asked me to make sure to tell you, tell Teresa that we don't believe it. We don't believe it, do you hear me? We know him. He wouldn't hurt anybody."

BY THE FOLLOWING DAY the wind had swept away the clouds. I was expecting another milky day, lingering rain, a landscape matching my dejection, but instead the sky was crystal clear and the rays of sun slanting through the countryside brought a new warmth. The first day of spring, a week early.

Outside the town's newsstand was a small post with a scrolling digital display in block letters: TRAGEDY IN A FAMILY OF SPEZIALE. So the news they'd missed about Nicola and Bern being related had emerged.

"Where are they talking about it?" I asked Maurizio, the newsagent.

"Everywhere. But especially here."

I glanced quickly at the headlines of the *Quotidiano di Puglia* and the *Gazzetta del Mezzogiorno*. Both front pages showed the same photograph of Nicola that I had found online the day before. I hunted for some coins at the bottom of my bag.

"Don't worry about it," Maurizio said, folding the newspapers.

"I don't see why," I said, handing him a fifty-euro bill. "That's all I have."

"You can pay me another time."

"I said no."

He took the change from the cash register. Meanwhile, other customers had gathered. I knew them, and they knew me. I noticed their looks, their eyes went from the newspaper headline to my face, then back to the headline. Maurizio very slowly counted out the bills. When he looked up, his expression had changed from before. He said, "When they were kids they used to come to the newsstand and stare at everything, their eyes popping out of their sockets. My father always told us about it."

In the car I read the article in the *Gazzetta* in a rush. It didn't say anything that I didn't already know, only that the search for the fugitives had been extended to all of Puglia. That word "fugitives" hit me. Pictures of Bern and Danco and Giuliana were provided, and everyone was urged to cooperate.

I noticed that Nicola's age was wrong, thirty-one instead of thirty-two. He had turned thirty-two the month before, on February 16. I had sent him a happy-birthday message and he had replied thanks with a lot of exclamation points. For years that's all we'd done, sent meaningless birthday greetings.

I looked for the obituary page. His item was the first. He was remembered by his parents and, in the box below, by his police colleagues. Nothing was said about a funeral. I went to the *Quotidiano di Puglia* and reread the same information, the same error about Nicola's age, but here they'd added that the funeral had been postponed because of the autopsy. When I looked up from the page I saw an elderly gentleman, one of the regulars in the piazza, seated on his bicycle, stopped a few steps away from the car. He was staring at me.

When I returned home I found the school bus parked in front of the masseria. The children were gathered around, each with his backpack containing a bag lunch. I had completely forgotten about the visit scheduled for that morning. The teacher, Elvira, and her colleague were waiting for me under the pergola, twisting their hands. I apologized for being late. It sounded ridiculous.

"We weren't sure if you'd feel up to it," Elvira said.

"It's fine."

"I'm sure everything will be cleared up, Teresa."

She touched my arm, gently, and that unexpected contact startled me. I turned to the children: "Did you find the goat? Yesterday I left her pen open. Go look for her, go on. She usually wanders off that way." I waved my hand to send them away and they started running in the direction I'd indicated.

Later I looked on as they carved pumpkins and scooped the orange pulp onto the ground. I distributed carrot seeds, one apiece, then watched them dig holes in the soil with their fingers, deposit the seeds, and cover them back up, full of hope. I promised that I would look after the seedlings, knowing full well that I wouldn't water the seeds even once, that I would let them die of thirst, every last one of them.

"Now do whatever you like," I said. "Run, climb, tear the place apart for all I care."

I went into the house without bothering to say goodbye to the teachers. I closed the door and sank onto the sofa. I was still there, perfectly alert, when the school bus drove off along the dirt track.

CONTRARY TO WHAT was originally theorized, Nicola had not died from the spade striking the back of his head. According to the autopsy, that blow had caused a fairly mild concussion. It had been the impact with a sharp rock that had caused much more severe internal hemorrhaging, an impact that the fall alone was not able to justify. "A further contributing factor must have crushed Belpanno's head forcefully against the rock," the press release reported. A further contributing factor. On the temple, on the opposite side, there were bruises compatible with the cleated sole of a massive shoe, a boot, maybe an army-issue rubber boot. Someone had flattened his head against the rock with his foot.

Just when the date of the funeral was announced, Danco's jeep was found along the coast, parked in a grassy clearing. The area was rarely frequented in the winter, the article online said, whereas in the summer it was always crowded because it was just a few steps from a well-known hangout for young people, the Scalo. Reading that, I felt dizzy. I saw Nicola and myself there many years ago, me unhappy with being alone with him, and him looking for an excuse to keep me there.

The investigators' opinion was that Bern, Danco, and Giuliana had fled by sea, with the help of an accomplice. It did not appear that any of them was familiar with navigation. Inside the ruined tower, a few hundred yards from the jeep, the carabinieri had found a bag with a few clothes and some food remains. According to the reporter, the escape was an implicit admission of the crime. And the hideout, as he insisted on calling it, showed it was premeditated.

The thought of Cesare haunted me. Should I send him a telegram? Was it already too late? There were lists of appropriate phrases on the internet; I read and reread them, but none seemed even remotely adequate.

"Our thoughts are with you . . ."

"May the eternal memory of . . ."

Finally I dropped the idea altogether.

I was undecided about the funeral as well, up until the last minute. An hour from when it was due to start I was still at the masseria, dressed in my work clothes, wandering around aimlessly in the hope that time would leap forward three hours, or maybe ten years. I had to make a mad dash along the highway, under a persistent, furious rain, straightening my hair with my fingers and trying to rub my eyes free of the expression of loss that had marked my face for days.

Police headquarters had insisted that Nicola be accorded a state funeral: a clear sign of solidarity toward law enforcement. The pews of Ostuni's cathedral were filled from the first to the last row, even the back of the church and the aisles were jammed with people standing: policemen with their families, carabinieri in ceremonial uniform, ordinary citizens drawn there by indignation.

I kept away from anyone who might recognize me, especially from Cesare and Floriana, who were virtually unreachable in any case, wedged between Nicola's coffin, covered with flowers, and the wall of people behind them.

I spotted Tommaso standing beside a column. He was trying to go unnoticed. There was just a glance between us, more hostile than emotional, our mutual distrust was still there.

The ceremony took place in utmost silence, as if we were being watched from above. The bishop called a younger priest, Don Valerio, to the pulpit. It was only after I'd heard him speak for a while—after he said, "I visited the house where Nicola lived with his parents a number of times, I blessed that house year after year"—that I remembered a muggy August day when Cesare had told me about his friend the priest in Locorotondo.

And here he was, Don Valerio, his furrowed forehead barely projecting above the lectern, his eyes dark and blazing. He described the masseria as a perfect sliver of the world, in which evil could not intrude. But evil, he said, had even managed to creep into the Garden of Eden in the form of a snake.

The bishop had sat down. He listened to the priest with his eyes closed. Don Valerio continued: "There is something that we cannot accept. Didn't the Lord promise us eternal life through our children? And now it seems He's revoked that promise. Cesare and Floriana would have a right to doubt God today, but I know that they will not. Because they have made faith the foundation of their every action. Listen well to what they have to teach us on this day of mourning, when even the heavens have joined in our weeping: every single moment of our stay on this earth makes sense as long as we believe in Jesus and in eternal life. If we stop believing, we might as well sit in a corner and let ourselves die."

He paused for a long moment. The bishop had bowed his head. I looked for Tommaso again, but he was gone. Don Valerio bent the microphone toward his mouth, but when he resumed his talk he was more subdued, as if his strength were about to run out: "I've been hearing a lot of talk these days. I hear accusations being made. As often happens, people talk without knowing what they're talking about. We all love gossip, don't we? And what's better than a violent death to generate gossip? Well, I saw Nicola with the boy he considered a brother. With Bernardo."

The name was a shock. The bodies crammed into the cathedral flinched.

"When I knew them, I saw two boys incapable of hurting anyone, let alone hurting each other. Raised with so much love as to make them immune to wickedness. I could be wrong, of course. As I've already told you, the snake even corrupted Adam and Eve. But let's wait for the moment of truth. This is not yet that time. This is the time of mourning and of prayer."

There was another talk after his, by a colleague of Nicola's, who unfolded a sheet of paper with trembling hands and read from it, stumbling over every word. He described Nicola so differently from how he was in reality that I lost the thread. For a while, after he returned to his seat, there was only the din of the rain hammering the roof, as the bishop blessed the wooden coffin.

That's when the scream erupted. An animal scream, which rose from a terrible depth, a scream that the local newscasts would air that night, the following day, and the day after that. Cesare held Floriana by the arms as she tried to lunge forward, not exactly toward the coffin, but toward something that she alone was able to see.

I elbowed my way through the paralyzed mob, not moving toward Floriana, but in the other direction, toward the exit blocked by the crowd.

There were people standing outside as well. I ducked under the umbrellas, pushing and shoving to make my way past them. The bishop had started talking again, the loudspeakers blared his voice: "Eternal rest grant unto him, O Lord . . ."

A hand grabbed me by the shoulder. I tried to shrug it off, but it tightened its hold. I turned around. Cosimo stared at me, distraught.

"What did you all do to him? What did you do to that poor boy?"

His blotchy red face was very close to mine. His white hair was sopping wet, the shoulder pads of his jacket sodden with rain, like sponges.

"I had nothing to do with it."

He was still clutching my shoulder. A lady nearby watched us but did not intervene.

"There is only hell for people like you!"

I managed to struggle free, or maybe he loosened his grip. When I made it out of the crowd, I too was drenched. My umbrella had been left in the church, but I didn't dream of going back for it. The rain had turned the pavement into a water chute. I fell once, twisting an ankle. Someone approached to help me, but I was already on my feet and running more precariously than before.

Driving back to the masseria, I tried to shut out the thoughts swarming through my head from the funeral: Floriana's animal cry, Don Valerio's words and those of Cosimo, the wreath of damp flowers lying on the coffin. The windshield wipers swept the glass at top speed, but that wasn't fast enough, it was pouring so hard, too hard, you couldn't even make out the road.

I REMEMBER very little of the weeks that followed. There was more rain, pounding at first, then intermittent, until there were only puddles left scattered in the yard, and finally those too dried up. Then came the inconsolable croaking of the frogs, all night long: I thought of the first summer with Bern.

April. A scrawl appeared on a wall along Speziale's main street: NICOLA LIVES. A few days later, LIVES had been covered over with an offensive word in red letters, and a circle of the same color had been drawn around the letter A, forming the symbol of the anarchists.

May. I lived as though suspended. A sirocco wind blew for weeks and already there was talk of the drought that would devastate the countryside in the months to come. That uncharacteristic spring, oppressive and dry, only intensified the feeling of stasis.

The police search had dug up traces from the past. I found a Bible that had belonged to Bern and the others. I spent a lot of time leafing through it. Comments in the margins, in tiny letters in three different handwritings, noted the meanings of the most difficult words:

> *stranger (a man from another country)*
> *diadem (a kind of necklace for the head)*
> *fetid (very smelly)*
> *grotto (cave)*
> *trickle (drip)*
> *caducous (destined to live a short time)*
> *halter (rope for horses)*
> *pervert (someone who has illicit and evil thoughts)*
> *scourge (catastrophe, grave disaster, often sent by God because of a*
> *   sin committed)*
> *drifter (someone who no longer has a place to stay and wanders around*
> *   the world like an exile, dejected and solitary)*

"Drifter." I repeated that word in a whisper. I constantly wondered where Bern was. Only his return could restore the normal flow of time and the seasons.

To keep me company there were the bugging devices. Truthfully, I hadn't found any, I hadn't even looked for them, but I knew they were there, that the police had planted them around the house. I also knew that the phone was being tapped, and that from time to time plainclothes agents drove up as far as the iron bar across the track, remained parked there awhile, and finally left. It made sense. All of that commotion on their part made sense. My husband was wanted for killing one of their own; an international arrest warrant had been issued against him.

Nevertheless, what the bugs recorded was irrelevant. Not just because Bern would not show up there and wouldn't even phone, but above all because the devices couldn't capture anything of what the masseria really was, of what it had been. They were listening for clues encoded in my conversations, interpreting sounds, but they couldn't pick up the countless happy moments, the years of our living there together, Bern and me, the mornings in bed and the long, leisurely meals, when we let ourselves be mesmerized by the rustling foliage of the pepper tree outside the window. They didn't capture the exhilaration of the years the six of us had lived there amid a glorious chaos, nor the intensity of our feelings for one another, at least at the beginning. And they did not capture the hope that had infused the masseria, since Cesare's time there. The only thing the hidden microphones could detect was an acoustic rendering of my loneliness. The rattling of dishes and cutlery. The gushing of the taps. The clacking of the computer keyboard and, in between times, the long hours of silence.

THE FIRST to appear on television was Giuliana's father. He said what I already knew, namely, that he had not been in touch with his daughter for ten years. But Danco and Giuliana were not all that interesting for the public. The fascination lay entirely in the blood bond between the cousins who were once inseparable and later became so hateful that one could kill the other. Bernardo and Nicola. Nicola and Bernardo. Simply mentioning their names as a pair was all it took for everyone in Italy to know whom you were talking about. Alternatively, all you had to do was bring up Speziale. A cloud of gossip rose around the masseria. Now that the incubator of that scandal had been disclosed, reporters and cameramen even ventured as far as the house. After turning them away at the door, I watched them moving about the property, looking for the best angle to photograph the dwelling. They wanted to take my picture as well, and a couple of them succeeded.

There were phone calls and emails sent to the masseria website, mostly from television stations, though sometimes they were purely obscene insults. My parents again tried to persuade me to return to Turin, just to find some peace while waiting for things to settle down.

At the newsstand in Speziale the covers of the weeklies featuring Bern and Nicola continued to be on view outside, displayed with perverse pride. I stopped walking past there, then I stopped going to town altogether. I did the shopping in supermarkets miles away, run by immigrants, and always during hours when they were deserted.

Just when the media found themselves running out of new developments, just when the focus on Bern and Nicola was finally tapering off, Floriana appeared in a television broadcast. It was aired early on a Wednesday evening and was watched by more than a million viewers.

Since there was no television at the masseria, I drove to San Vito dei Normanni that evening, where no one knew me. The one-way streets were clogged with cars. I passed a bar and through the window I saw a monitor hanging on the wall. I parked. There were only men inside, with the exception of the bartender, a corpulent woman with a clinging yellow tank top and a tattoo on her arm. They studied me silently as I passed between the tables.

I sat down in a spot closest to the screen, turning my back to everyone. I ordered coffee and didn't notice when it was put on the table because Floriana had already appeared in the video. Behind her, a kitchen stove I had never seen before.

She nodded in response to the interviewer's welcome, and the interviewer then opened by saying, "Perhaps many viewers may not remember Floriana, but I do. For the women of my generation, little more than twenty-year-olds at the end of the seventies, she is a symbol. Floriana

Ligorio was among the first women to oppose the detestable practice of exploitation of farm laborers in the region she comes from, Puglia."

Floriana nodded without answering, because that wasn't a question. The host went on, addressing the public directly now: "There is a photograph that became famous at the time. Here it is, the girl whom the policeman is holding by the arm is Floriana."

For a few seconds the image filled the screen. Soon afterward it appeared smaller, lying on the table between the two women. Floriana looked at it without touching it, as if she doubted it was really her.

"Do we sometimes have to struggle to get what's right, Floriana? Struggle even against a policeman?"

"He was gripping my arm, all I tried to do was free myself."

"In an interview at that time you called the policeman in the photograph a 'dirtbag.'"

"Our battle was a just one."

"What would you think if your son, Nicola, in a similar picture taken today, were the policeman gripping the girl's arm?"

Floriana's head snapped up. "He wouldn't have done that."

"Can you describe your relationship with your son?"

"On Sundays he came to see us. For lunch. If he wasn't on duty."

"And you never had any arguments? After all, it seems clear that your ideas differed. You were a symbol of dissidence, and Nicola became a police agent."

"A mother is able to accept her child's choices."

The bartender in the yellow tank top approached the table and asked me if there was something wrong with the coffee. I said everything was fine.

"You didn't drink it," she said, grabbing the cup.

On the screen, the interviewer had noticed Floriana's agitation, an agitation that she herself had provoked, and was now reassuring her that everyone, she herself included, was on her side, and shared the pain of her loss, a cruel loss.

"But here we are, and this is a unique opportunity to shed some light on every aspect of this affair. We have to face it bravely, Floriana. There are many testimonies from demonstrators who were at the scene of the murder. They witnessed Nicola and his colleague's arrival. They tell of an aggressive, defiant attitude. Some claim to have been verbally attacked, and others say that Nicola continued fingering the belt holding the gun in an incendiary way."

At that point Floriana lost control: "It's my son who was killed. My son, Nicola, was murdered by a group of terrorists. He's the one who's dead! That's what we should be talking about!"

"Is that what you consider them? Terrorists?"

"What else would you call them?"

The interviewer nodded, then gave a summary of what had happened. She reminded viewers that the principal suspects in the death of Nicola Belpanno were Danco Viglione and Bernardo Corianò, the victim's cousin. Their photographs were shown. She asked Floriana if there was anything in Bern's past, some episode perhaps, that would give any hint of the violent young man he would become.

"He had his odd ways, like all children. He grew up without parents."

"Do you mean that Bernardo is an orphan?"

"Cesare's sister, Marina . . ."

"Cesare is your husband, right?"

"Yes."

"So we're talking about your sister-in-law. We're just trying to make it clearer for those who are watching us, Floriana. Continue, please."

"Marina was very young when she became pregnant, she was only fifteen."

"Fifteen?"

"She came to us because she didn't know where else to go. If she had told them at home . . . My in-laws were very strict people. Nicola had just been born and we had bought this little house in the countryside and fixed it up. There wasn't even a well. We had to go to the village fountain every day and fill our containers with water."

"Were you hippies?"

"No. I mean, we didn't see ourselves that way. Hippies don't believe in God."

"Whereas you and your husband are very devoted."

"Yes."

"Your husband even founded a sect."

"He wouldn't want it to be called that."

"Let's go back to Marina, your sister-in-law. She came to you to ask for help because she was pregnant. And so young. Southern Italy, in those days . . . It must not have been easy."

"She wanted to find a solution."

"What kind of solution?"

"She was only fifteen, she was scared."

"Do you mean she wanted to have an abortion?"

"Cesare was very concerned about her. He's the big brother, he's ten years older, and in their family . . . he's always been like a father to Marina. But he too was very young. We were all very young and we had no money. One evening, Cesare went off somewhere. He stayed out all night and when he came back he said that we would take care of Marina's child no matter what."

"What had he done that night when he stayed out?"

"He had prayed."

"Did your husband often do that? Stay out all night praying?"

"Sometimes he did."

"So we can say that Bernardo was born as a result of your husband's praying."

"Yes."

"And can we say that with his praying your husband saved the child who thirty years later would kill his own son?"

Floriana fingered the transparent frame of her glasses, as if to make sure they were still there. A few seconds of silence followed. The interviewer shuffled her sheets of paper.

"All right. We'll continue with this, Floriana. We'll pick up exactly where we left off."

There was a commercial break. When the program resumed, it was not a continuation of the interview with Floriana. Instead, images of Speziale's town center were shown: the street that cut the town in two, the bar-café, the grocery store, and the small church where many years ago I had attended my grandmother's funeral mass.

A car was driving along country roads that I knew all too well, with dry stone walls on either side of them. Taking the longest possible route, it arrived at the barred entrance to the masseria. The reporter had no qualms about ducking under the iron bar and walking along the dirt track, on my property, followed by the cameraman. He went as far as the house, where the doors and windows were closed.

The interviewer said: "Your husband preferred to see to Nicola and Bernardo's schooling himself. Why?"

"Cesare is an educated man."

"There are many educated people who, nevertheless, decide to send their children to school."

"We had our convictions. We still have them."

"Meaning you would do it all the same way again?"

"Yes. Or maybe not everything. Not all of it."

"In view of what has happened, don't you think that the isolation may have contributed to altering Bernardo's personality?"

"Bern. We always called him Bern. No one calls him Bernardo."

"Bern, of course, excuse me."

"Cesare gave all of them an excellent education. Better than that of other children their age."

"Is it true he forced them to memorize passages from the Bible?"

"No, that's not so."

"When we spoke the first time you told me that . . ."

"I said that Bern memorized parts of the Bible. He was the one who wanted to. Cesare never forced him to. For him, they only had to know a few short passages, only what was essential."

"Essential for what?"

"For them to understand."

"For them to understand what, Floriana?"

The camera panned for a moment to the interviewer, who was now frowning.

"Floriana, I think it's important that you clarify this point. What did the boys have to know at all costs?"

"The principles of faith. The principles of . . . behavior."

"And were there punishments for anyone who refused to learn what Cesare considered essential?"

Floriana shook her head slightly, as if shivering.

"During our first conversation you told me that there were severe consequences for those who did not obey him."

Floriana was silent. The interviewer lowered her voice even further.

"Did his son, Nicola, and Bernardo ever receive corporal punishment from Cesare?"

Floriana turned her head purposefully, as if looking for someone. Then there was an abrupt cut. In the next frame she had a half-full glass of water beside her. Her upper lip was a little moist. The interviewer was more severe than before.

"After Nicola's death you decided to leave your husband. Do you think that he is partially responsible for what happened?"

Floriana took a sip of water. She stared at the glass with a lifeless expression. Then she nodded.

"Why did you decide to share your story today?"

"Because I want people to know the truth."

"Does the truth set us free, Floriana?"

Floriana hesitated, as if the question had triggered a startling memory. Her eyes widened a little more, just for a moment. With conviction she said: "Yes, I think so."

"What would you say to Bernardo if you knew he was watching us right now?"

"I would tell him to stand up to his responsibilities. As he was taught to do."

"And what would you say to Nicola if you could, Floriana? What would you say to your son?"

"I . . ."

"Would you bring Mrs. Belpanno some tissues? Don't worry. Take all the time you need. Have some more water. Do you feel up to continuing? We were talking about Nicola. When I came to see you, you told me that Cesare has his own very personal idea of the Catholic religion. He is convinced that souls are reincarnated. So, if you were to close your eyes and . . . Close your eyes, Floriana. Try to imagine: into what creature would your son have been transformed?"

The picture on the screen suddenly changed, as Floriana's mouth opened. A music show appeared on the monitor.

I went to the counter, but the bartender paid no attention to me; she was busy listening to another story from the guy with the paint-spattered face. When I interrupted them, they both turned to look at me.

"Please put the channel with the interview back on."

"Nobody's interested in it."

"I am."

"For just coffee you can go watch television at home," the barmaid snapped, then turned back to the man.

He didn't say anything. He brought the beer bottle to his lips and took a swig, still staring at me.

"I'll have a beer too," I said. "It's important, please."

She picked up the remote control from the counter and, instead of changing the channel, placed it on the shelf behind her, in front of the bottles.

"The coffee is on the house. Now get out of here. We can tell you're a friend of those people."

I left. Stunned, I wandered through the streets of San Vito, now deserted. I didn't find another bar, just a kiosk selling watermelons, with a small TV set up top, but I took a look at the person sitting there and didn't dare approach. I thought of ringing a doorbell at random, but I had done enough foolish things; I was tired, drained.

I did not see the end of the interview. Only many months later did I learn how Floriana had answered the question: she said that she had never really believed in reincarnation, that for years she'd let her husband mislead her, but not anymore. She said that the Lord grants us only one life on earth, this one, and there won't be another one afterward.

ONE AFTERNOON in August I was keeping an eye on one of the intruders through the bedroom shutters. He had no camera. After knocking, he'd wandered around the yard for a while. I'd lost sight of him as he walked around the house, then he'd reappeared and looked purposefully at my window, as if he sensed my presence. He'd sat down at the table under the pergola, half hidden by the dense tangle of vines, and he hadn't moved from there.

Half an hour went by, maybe more, and he still showed no signs of leaving. Suddenly I was enraged. I went downstairs and opened the door like a fury.

"Go away right now!" I shouted. "You can't stay here!"

The intruder leaped up. For a moment he seemed about to obey me, but then he stayed where he was.

"Are you Teresa?"

He was younger than me, overweight, and he looked harmless. He wore battered Birkenstock sandals and a sweat-stained T-shirt.

"You have to leave immediately," I repeated, "or I'll call the police."

But instead of going away, he seemed to get up his nerve. He took a step toward me and bobbed his head in a kind of bow.

"My name is Daniele. I'm a friend of his."

"A friend of whom?"

"Of Bern. He—"

I put a hand over his mouth and gestured for him to follow me to the olive grove. When we were a reasonable distance from the house, I inundated him with questions; Daniele answered me patiently, as if that outburst was just what he expected from me. He told me that he had met Bern at the encampment in Oria, more than a year ago, and from that moment on he had always been by his side. He was also with him on the night at the Relais dei Saraceni, but he hadn't seen what happened. As he spoke, he avoided meeting my eyes. He addressed a spot somewhere to my right, from time to time wiping his flushed face with his hand.

"Can we move out of the sun?" he asked at one point.

I realized then that I had led him into the middle of a sunlit clearing, its soil scorched, as if even the trees were infested with microphones.

We stepped under an olive tree. He was panting a little. I asked him why he had waited so long before coming to find me.

"I was under house arrest. For four months. The police thought I was the one responsible for the arsenal, just because I'm studying chemistry. But they had no proof. And studying chemistry is not yet a crime."

"Was it true?"

"What?"

"Were you responsible for the arsenal?"

Saying that word, "arsenal," seemed a little ridiculous to me. Daniele shrugged.

"Anyone would be able to make those explosives. On the internet there are thousands of tutorials."

He looked around, squinting in the direction of the house, as if looking for it over the barrier of trees, then he whirled around.

"The food forest is that way, right?"

"How do you know about it?"

"He always talked about this place. About the masseria. He described every last detail to us. And over there is where the bees were kept. Where the reed bed is."

Hearing him mention the reed bed made me feel a little dizzy.

"I'm saving up some money," he went on, not noticing. "When I have enough I'm going to buy some land. In fact, to be honest, I already found it. There's only a ruin there for now, but it can be fixed up. It will be like the masseria."

"Would you like to see the food forest?" I asked him.

His eyes lit up. "Will you take me there?"

But as we walked among the plants that were wilting in the sultry dog days, I got the impression that Daniele had already been there. Probably he had looked for it on his own, while he was waiting for me to come out of the house.

"Do you really never give them water? Not even a drop?"

"During this period, yes. A couple of times a week."

He bent down on his knees to examine the wall of branches on which the aromatic herbs grew. He touched it.

"It's just like he described it to us."

I asked him how old he was. "Twenty-one," he replied.

He stood up. "I had never met anyone like him. He was a great inspiration for me."

"Take me there," I said.

I hadn't planned to ask him. I hadn't even thought I'd want to. Daniele looked at me. "Where?"

"Where it happened." He shook his head. "I really shouldn't be seen around there."

There was a moment of silence. Then he added, "If you want, we can go to the encampment. In Oria."

"The encampment was disbanded."

He looked around. "Do you have anything better to do?"

"AND ANYWAY, the encampment can't be disbanded," he said once we were in the car. "They can drive us out, but we continue to exist. It's Bern who taught us that. We already have a place near Tricase, an old limestone quarry. But we're waiting for things to settle down."

You could barely see the road through the dirt-streaked windshield. Daniele was bent forward, clutching the steering wheel nervously. With his right hand he yanked the gear shift.

"Many of us are still under surveillance. One guy even had a plainclothes cop come to the university. The organic chemistry professor asked the agent if he was new. Then he wrote a synthesis on the blackboard and asked him if he understood it. The cop turned purple with embarrassment, and practically fled the classroom. He had even brought a notebook to take notes."

The car jolted as he changed gear. He cleared his throat.

"But we're in the South, nothing works really efficiently for too long. Pretty soon even the plainclothes agents will give up and go away."

A heavy-metal piece blared from the car radio. From time to time Daniele began shaking his head in time to the beat, lip-synching the words. Then he asked me, "What do you know about Oria?"

I looked out the window, a little embarrassed. Nothing, I didn't know anything.

"There were about forty of us," he explained. "We made the rounds night and day, divided into groups, it was exhausting. But the area was still too big to patrol all of it. Hundreds of olive trees marked with a red X, can you imagine? Danco drew up very complicated work-shift schedules with times and routes. If a group ran into one of the cooperatives hired to fell the trees, someone would have to split off and run in search of reinforcements. The remaining ones were too few to protect the olive trees. Not to mention the fact that the tree cutters often showed up at several places at the same time. In a word, they were trouncing us."

"You talk about it as if it were a war."

Daniele turned around. "What else would you call it?"

The roadside was littered with garbage; beyond it stretched tomato fields and olive groves, a line of purple haze on the horizon.

"Danco's strategy couldn't work. He proposed solutions that were more and more far-fetched. He sent the younger guys around to measure the distances between the olive trees, saying that with precise mapping he would be able to monitor everything. And in the meantime the Xylella epidemic kept spreading. We had reached a dead end."

When Oria was already visible in the distance, we took a dirt road. Instinctively, I checked the phone: there was no signal. Daniele stomped heavily on the pedals with his sandals, again silently mouthing the words to the song. I had gotten into the car of a stranger, I had let him take me to a remote spot in the countryside from which I would not be able to call for help, solely because he had claimed to be a friend of Bern's. I asked him if we were nearly there, and he nodded without turning his head.

"Bern didn't draw attention to himself much," he resumed after a while. "He stayed in Danco's shadow. I'd hardly noticed him. It seems impossible, but it's true."

We got out of the car in the middle of nowhere, crossed a mowed wheat field, and found ourselves at the edge of what must once have been an olive grove. But all that remained of the trees were the flat stumps of their trunks, and piles of branches and dry leaves. There was no trace of the encampment anymore.

Daniele went on talking: "One day we were all feeling very disheartened. We sat cross-legged, in silence. Bern stood up. He started walking around. He went from one olive tree to another, as if he were listening to an inner voice that told him now this way, turn around, another ten steps. Then I saw him grab on to the main limb of an old, majestic olive tree."

Daniele turned to look at me. He smiled. He pointed to a stump about twenty yards from us: "That one."

We went over to it. He touched the flat part of the stump, circled one of the rings with his finger. I felt like doing it too, but something about sharing his emotions about Bern embarrassed me.

"Can you picture it?" he asked. "It was very tall. Bern climbed up. We watched him scale the upper branches, so high that he became invisible. He stayed there for the rest of the day. A number of people tried to persuade him to come down, but he didn't pay any attention to them. I only approached the following morning. Nothing had changed much, in any case. Bern was still in the tree, just in a different spot, which to him had seemed a more comfortable place to spend the night. By then we had all moved over there. And that was maybe the first lesson that Bern wanted to teach us: that one single symbolic act is more powerful than a thousand obvious, recurrent actions. But I say that today. There are a lot of things that I only now understand."

Daniele was silent for a few seconds, as if he were doing so that very instant: understanding something that had previously escaped him.

"After a couple of days, that was all we talked about. Bern would not come down from the tree and we had to devise a makeshift pulley to send up his meals and water, his toothbrush and toothpaste. One evening the temperature plunged; at dawn the tents were all soaked with dew. Bern asked for a sleeping bag and someone told him that he could very well come and get it himself. The situation had begun to get on people's nerves. His stance seemed to ridicule the whole initiative. Even Danco kept away from the olive tree; for the most part he stayed in his tent, drawing up increasingly complicated shift schedules, strategies for patrolling and remote communication. But not me. I sensed the power of Bern's act; I felt it in my bones. So I brought him a sleeping bag. I rolled it up, crammed it inside a backpack, and clambered up. 'Not on this branch,' he told me. I remember it perfectly: 'Not on this branch, it won't support us both,' as if the olive tree were a grandiose extension of his body and he knew its strength, the way each of us knows the strength of his own arms and fingers. I left the sleeping bag where he'd indicated. For a while we stayed there, gazing at the expanse of olive trees that stretched out beneath us, not saying anything. To Bern I seemed no more relevant than the birds that perched and then flew off. There was something in his eyes. A determination. A flame. The others were making supper, and from up there their bustling about seemed so insignificant. Then Bern said, 'Tomorrow I'd like a bucket and some soap.' Not 'Please,' not 'Do you think you can bring me?' And to make it clear that he didn't want me up there with him a second time, he specified: 'You can use the pulley.'"

Daniele had sat down on the stump. When he turned to smile at me again, I saw that he was moved. So I sat down next to him. The wood seemed to transmit a strange warmth.

"I became his on-the-ground aide. His assistant, that's right, the others acknowledge it today. That I was the first to put my trust in him. Without my dedication, Bern wouldn't have made it. They would have starved him until they forced him to come down, or he would have let himself starve to death, who can say? In any event, we wouldn't be who we are today. But I'm not sure I deserve all the credit. It may seem absurd, but I'm convinced that it was he who chose me. Every morning I loaded what he needed on the pulley and sent it up. When the rope was yanked twice I brought the basket back down. Inside was a list for the following day. He washed his clothes himself, in the bucket, and spread them out to dry on the more slender branches. During the day he sat motionless for many hours, and at night he disappeared inside the sleeping bag. Not even the rain was able to discourage him. At first he let it fall onto his head. Then he sent me a request for a ball of twine and some scissors. It was the only time he added 'please.' I couldn't find any twine around, and I became agitated. By then I never even considered the idea that Bern could come down and take cover. It would have been a huge betrayal. It rained harder and harder. Finally I took the fasteners from my tent, letting it sag under the pelting torrents. I sent the cords up to Bern. I stayed and spied on him from under the dripping branches, the fat drops hitting me right in the eye. I saw him break off a few carefully selected branches and weave them together using a very complicated technique. Within an hour he had built a roof that at least partially covered the sleeping bag, a roof resistant enough so that the water was diverted to one side and fell into a single teeming cascade. He went underneath, crawled inside the sleeping bag, and did not send down any other requests."

We stood up and started walking again, a random zigzag course among the olive-tree stumps. Daniele was perspiring heavily, and by then so was I.

"They came in the dead of night," he went on. "It was the best way to catch us off guard. They surprised us from several directions at the same time. It was clear from the outset that the intention wasn't only to chop down as many trees as possible, but to disband the encampment once and for all. The carabinieri were there; it was a full-scale ambush. We shot out of the tents. We knew what to do, Danco had instructed us, so we scattered into small groups, one per tree, in strategic positions. 'The maximum possible coverage,' Danco always said. It was freezing, the ground was damp, and we were barefoot, some in undershorts and T-shirts, yet we took our places according to plan, each group to its olive tree, our backs pressed against the trunk, holding hands. From one post to the other we shouted insults at them and words of encouragement for us, trying to be heard over the sirens and engines and the roar of the buzz saws that were then switched on, as if the tree cutters were ready to slash any obstacle in their way, us too, if necessary. The carabinieri managed to disperse the first groups. There were some very young boys, still minors, all it took was a threat to call their parents. I was paired with Emma, one of the first to form the encampment. Her hands were frozen, her lips purple, but she was so furious she paid no attention. I was afraid that if I loosened my grip, she would run up against one of the tractors and try to stop it by kicking and punching it. The truth is that we were impotent, so I held her hands and she struggled, because all we could do was protect that one single olive tree. Have you ever heard the crack that a centuries-old tree produces when it crashes to the ground? It's not a crack, it's an explosion. The ground shuddered. And suddenly I remembered Bern, Bern on top of the olive tree that I could glimpse on and off in the blue gleam of the flashing lights. I was sure he hadn't moved from where he was. The trunk of his tree was assigned to Danco and Giuliana. I felt the urge to run to them and reinforce its defense, but I couldn't abandon my post, it wasn't in the plan. Then the tree cutters switched off the electric saws and backed away a few dozen yards. The carabinieri were wearing gas masks and they started hurling tear gas. As a result they managed to drive all of us away from the trunks. The order they had received must have been clear: no more wavering. They allowed us to get our blankets and clothes from the tents, and we searched for them blindly, our eyes burning. Those who struggled the most were handcuffed, Emma as well. It wasn't necessary for me or for Danco, we knew it was useless to get worked up. We watched the tree cutters with their safety helmets and reflector coveralls attend to one olive tree at a time, very calmly now, with a kind of satisfaction."

Daniele spread his arms.

"Imagine a grove," he said. "Here. Trees everywhere. Now look at this. A few days later Bern told me how it had been to watch from above, to see the crowns of the trees crumple one by one. He said he'd cried at first, but at a certain point he'd stopped, and the sadness had been instantly replaced by rage. From up there, he said, you couldn't see the men who were sawing the trunks. You could only see the crowns disappear, as if something invisible were swallowing them up. It took many hours, all night. In the end only one tree was left standing. The carabinieri had spotted Bern through the leafy branches, with his clothes spread out to dry, the roof to shelter him from the rain and the system of pulleys and buckets. They wanted to keep that tree for the grand finale. It was almost noon when they returned to him. They ordered him to come down, otherwise they would climb up and get him and they would arrest him. He didn't answer. Not once did he answer, as if he didn't speak their language. The carabinieri conferred about who should go up and finally two of them clambered up, led by a very agile tree cutter who acted as trailblazer. When they were fairly close to Bern, he moved. He climbed higher, very nimbly, like a spider. And when the three men had almost reached the limb on which he waited for them, he started moving toward the end of it, clinging to it with his hands and knees, gripping it between his legs as it got thinner and thinner. The wood bowed a little. The tip was so slight that it wouldn't even support a child. Bern looked at his pursuers, still without a word, but what he was saying to them was very clear: one more step and this branch will crack. We were all standing there. Someone started applauding, and we repeated his name, 'Bern, Bern, Bern.'

"The carabinieri were nervous now. Those below ordered the ones in the tree to make him go back, otherwise he'd kill himself. The two in the tree repeated the order to Bern, but without conviction, by then they were scared. They backed up very carefully, trying not to disturb a single leaf. Bern did not move from the very tenuous branch tip until the second carabiniere also set foot on the ground. He had won. We had won. Most people were distracted, hugging one another, but not me, I kept watching him, so I saw it when he released his hold to move his hand forward a little and the branch suddenly gave way, as if the tree itself had been resisting with him the whole time, but now had no strength left. Rather than let himself plunge into space, he managed to hang on, but he struck his shoulder against the trunk. Maybe it would have started all over again, the carabinieri would have climbed up and he would have fled to another branch, and maybe the third or fourth time he would have had a serious fall. But his shoulder hurt and in any case he couldn't hold out forever. It was another thing he would later explain to us, as we waited for the X-ray report at the emergency room in Manduria. Still, when I saw him come down, I was disappointed."

The sun was disappearing completely behind us. The wind died down abruptly. It seemed as if the entire countryside had fallen silent to hear the end of Daniele's story.

"The tree was cut down. There was a hush. Danco went over to Bern and put an arm around his waist. Together they contemplated what had maybe been a defeat and what had maybe been a victory. We went to the hospital in Manduria and came back with Bern bandaged up and a supply of painkillers. He treated me as if he barely knew me, cordially, yes, but as if I hadn't been the one attending him the whole time he was in the tree. It hurt a little. Then he decided to leave to spend a few days at a friend's house in Taranto. When he returned here, to the camp that we still hadn't had the resolution to move, he brought the news about the tree felling planned at the Relais dei Saraceni. He described a magnificent stand of olive trees. He knew they weren't really infected. Then he said that this time it would be different."

"Different how?" I asked.

"He said he had realized that resistance was no longer enough. 'From now on we'll talk about fighting,' he said, 'and every battle needs weapons, even if that offends you, even if it's not what you envisioned. But look around you. Look what they did!' And while we all tried to take in what for the moment were merely ideas, shocking, sure, but just ideas, Bern started singing. On the tree stump, in front of all of us."

THE CLASS VISITS to the masseria did not resume with the beginning of the school year. I phoned the teacher, Elvira, but she didn't return my call. I tried again a few hours later, then the following day and the day after that. When she finally took the call, I could not contain my irritation.

"I wanted to know the fall calendar," I said.

"I'm sorry, Teresa. There are no visits scheduled."

"I'm planning to build an aviary, to hold all the species in the area. Redstarts, black-throated wheatears, thrushes."

The idea of the aviary had only come to me an hour earlier. I wouldn't even know where to start if I had to build one. Was it conceivable that birds of different species might live in the same cage?

"I'm sure the children will like it," I persisted.

"The teachers' council decided not to include the masseria among this year's extracurricular activities. I'm sorry."

"The last visit was disappointing, I know. The masseria was a shambles."

Would I have begged her? And would begging have changed anything? But Elvira did not give me time to find out. Her tone of voice suddenly changed: "How do you think I could justify it to the parents, visiting the house of a . . ." She stopped, she couldn't manage to say the word.

Then, as if expressing a resentment she'd harbored all summer, she added: "You should have told me, Teresa."

"Told you what?"

"I brought children there, for God's sake! Children!"

After that phone call I saw the masseria with different eyes. The gutter had broken off during a storm several months before, the viburnums had died of thirst, and the food forest was a wreck. As for me, I looked like only a distant relation to the woman in overalls who on the website was smiling and holding a garland of flowers.

I counted the money left in the tea canister: forty-two euros. That figure made me burst out laughing.

Even the gardener lost patience with not being paid. He came for the last time. He confirmed that the holly oak was infected and showed me the streaks of resin along the trunk, like bleeding wounds. It would die slowly; it could take years. Did I want him to come back and chop it down? I said it was fine like that. I chose to have the holly oak stay where it was, dying slowly, day after day, with me.

I sold the hens first. Then the beehives. I sold some tools and finally I sold the goat to a cooperative in Latiano. They wouldn't slaughter her because she was too old, but she could be impregnated and the kids would be born in time for Easter. They asked me if I wanted one, they would set it aside for me. I said no, thank you.

The drought of the last few months had caused an abundance of figs to ripen. I picked through the masseria's trees, then collected the ones on my grandmother's land as well. Riccardo had left a couple of weeks ago, he wouldn't know. I arranged the best fruits neatly in boxes, I tried to make them look appealing, putting leaves underneath and on the sides.

With the others I made some jams. The next day I loaded the car and drove to the traffic circle between Ostuni and Ceglie. I parked in the wayside, set the goods out on the lowered tailgate, and sat on the wall in the shade.

And so I'd become an itinerant vendor, like the bare-chested old men who peddled watermelons along rural roads. When I saw them, as a child, I wondered how they could make a living, and I often forced my father to pull over and buy something. He tried to explain to me that they were farmers, not paupers, but he couldn't change my mind.

A number of people slowed down to take a look at the boxes of figs, but they rarely stopped. The whole countryside was full of figs, and the tourists, the only ones who would have been impressed, had already gone. One guy lowered the car window and from behind his sunglasses, in a perfectly cordial tone, made me an obscene proposal. I recognized a farmer from Speziale with his wife, in an Ape 50, and they recognized me. They owned land in the area where the masseria was. They drove on without a nod.

Nevertheless, before sunset I had sold all the boxes and most of the jams. I would have to come up with something else soon, but in the meantime, by keeping my needs to a minimum, I could get by for a few weeks.

One evening, from a distance, I saw the neighbor's Ape 50 stop in front of the iron bar. He got out, deposited something on the ground, and drove off. I walked along the dirt track that was overgrown with weeds. I looked in the basket: fresh vegetables, two packets of pasta, a bottle of oil, and one of wine. Charity. I would never fully understand that place. I would never understand Speziale's rules or its inhabitants, their constant vacillation between hatred and compassion, their angry ways and their equally brusque acts of kindness. I cooked the vegetables and set the table under the pergola. It was the first meal I'd sat down to eat for months.

THE CHANGES MADE to the masseria the year before had at least left me with an internet connection; it worked sporadically, but the speed was acceptable. In any case, I wouldn't have wanted it to be any faster than it was: the seconds spent waiting for a new page to load, watching a blank screen, were moments of pure absence, without pain.

I mostly watched YouTube videos. I started from any one of them, then I let myself be shuttled along from one connection to another, no more decisions, lulled by that ephemeral geography. Outside, the sun went down, and the house sank into darkness, except for the bright rectangle of the screen where I was straining my eyes. I kept at it for quite a while, until very late. When I stopped, I'd be so groggy that I would drag myself to the couch to sleep.

One morning I was awakened by the rumble of wheels on the dirt track. I hadn't yet opened the shutters, though it must have been late morning. I remained lying there, staring at the grid of light on the ceiling. I heard a knock, then another a few seconds later.

"Delivery!" a male voice called out.

I went over to the window and opened the shutters just enough to peer down below, my eyes squinting in the sunlight. The delivery boy waved a package in my direction. "Are you Mrs. Gasparro? You have to sign."

I pulled on one of Bern's old T-shirts and went downstairs.

"I'm sorry I woke you."

"I was awake. I have a touch of flu," I lied.

"This district doesn't have a name, does it? I had a hard time finding it."

The package had the Amazon logo printed on it.

"What is it?"

"You should know, you ordered it," the boy said, smiling.

"I didn't order anything. I don't even have an account with Amazon."

He shrugged and held out a screen on which to sign.

"Use your finger. It doesn't matter if the signature isn't perfect," he said. Putting the device back in a side pocket of his pants, he added, "It must be a gift, then. Usually there's a card inside."

When I was alone, I opened the package. It contained a bottle, with a picture of a plant and some kind of magnified bug on the label. It was clearly a gardening product, but the instructions were in German and other languages that I couldn't even identify.

It had to be a mistake. In any case, I had nothing better to do, so I sat down at the computer and patiently typed the German words into Google Translate. The result was barely comprehensible, but it was enough to confirm that it was a natural pesticide. You were supposed to dilute a capful in ten quarts of water and use it to water the diseased plant every other night. In the end the gardener had felt sorry for me, or for the holly oak. More charity. I gave the tree its first dose of medicine and it was enough to make me feel better.

ONE DAY Daniele returned to the masseria. We sat under the pergola for a long time, sipping the carob liqueur he had brought. When we got up the bottle was empty.

Drunk as I was, I took him into the house, up the stairs, to my bedroom and Bern's. He let himself be led. I watched him undress in front of the bed, barely able to balance on one leg and then the other to pull off his socks. He had a flabby belly, it made me giggle.

Afterward I don't know what got into me. I licked his face and bit his shoulders until he begged me to stop because I was hurting him. Then I lost the will altogether, overwhelmed by a wave of sadness. I fell back on the bed, and in an instant I was far away from there. I let him do what he had to do, until it was over. The objects in the room expanded and contracted as they used to do when I was a young girl.

Only later did I remember the bugs. I wondered what the cops thought as they listened, how they judged the wife of the terrorist who, after having seduced a younger man, had talked to him for an hour about the husband who had disappeared, confessing how she missed his body, how she'd been missing him just before. And who knows how they judged the young man who had heard that confession without interrupting her, who'd never once stopped stroking her hair.

In the morning I woke up alone. Daniele was in the kitchen, where he'd made breakfast. The glasses we'd used the evening before were washed and turned upside down beside the sink. We ate in silence and right after that I told him to leave and not come back. He did not ask me to explain.

THE SECOND DELIVERY arrived a few weeks later, in October. The same van parked crookedly near the barren vegetable garden, the same delivery boy.

"So you finally signed up," he said, handing me the package.

"Signed up?"

"With Amazon. Don't tell me it's another mistake, because I don't believe it."

"Nevertheless, I'm afraid it is."

"The thing is that Amazon doesn't make mistakes. Are you sure you didn't order it?"

I signed the little screen with my finger.

"If I were you, I'd take a look at my credit card account," he said, "just to be safe."

This time I didn't wait for him to leave before unwrapping the parcel.

"Is there a card?" he asked, as I stared, dumbfounded, at the cover of the book.

Then he must have left, surely he left at some point, though I can't say exactly when because I don't remember anything about those minutes, except that I was alone again, still under the pergola, still holding the book that was shaking in my hands. I couldn't stop looking at it, yet I couldn't so much as leaf through the pages.

It was a different edition than the one I knew, more vivid, glossier, yet it was the same book that I had unsuccessfully tried to read many years ago, the same author, the same title: Italo Calvino, *The Baron in the Trees*.

I sat down at the computer and went to the Amazon site. I typed in my email address, and kept getting it wrong because my fingers were trembling. I followed the procedure to reset my password, a password that I had never chosen. A code was sent to my email, which was clogged with unread messages, advertisements, discount offers, and grotesque sexual proposals.

I entered the code, then I had to choose a new password. I stared at the screen for quite some time, my mind a complete blank, unable to decide on a sequence of letters and numbers. When I finally did, I found myself in my Amazon account, an account that I had never opened.

I clicked on "Your Orders" and two items appeared: the organic pesticide and *The Baron in the Trees*. I selected the book and a new screen informed me that I had purchased it on October 16, 2010.

There had to be a section on payments, but it took me a while to find it. The registered credit card coincided with mine, I recognized the last four digits, which were visible.

Increasingly bewildered, I left the house and drove to the only ATM in Speziale. During the summer it had been uprooted with a tow hook, but they had now replaced it with a new one. My current account balance was no longer in the red, though I myself had not deposited anything. I printed out the bank statement. It showed a wire transfer of one thousand euros made a month earlier, then the two Amazon charges and the card fees. The wire transfer had been made by my father.

I went back to the masseria. The business card must still be somewhere on the shelf. In the state of neglect in which I'd sunk, I'd done nothing but let things pile up: receipts, advertising flyers, empty food containers, balled-up plastic bags. I rummaged through the disorder, haphazardly at first, rooting around with my fingers, then tossing everything that didn't interest me on the floor. I found the card: "Alessandro Breglio— Computer Assistance." I dialed the number but stopped before the last digit. Someone might be listening.

An hour later I entered the store in Brindisi, cluttered with monitors and keyboards waiting for repair. The young man looked at me, trying to place me in his memory, but I didn't give him time.

"Is it possible that someone could have gotten into my computer and done things in my place? Such as buying things?"

His eyes lit up. "They hacked you? It's possible, sure!"

"And where can they do it from?"

He smiled. "Even from the moon. I can come and take a look if you want. We have a very reasonable security app."

"I need to track down the person who made the purchases."

"I think that may be very difficult. You can try the police, but from experience I can tell you that they won't pay any attention to you. Have a seat, let me explain how our app works."

"I don't want the app!" I may have screamed at him, because he shrank back in his chair, looking shocked.

After a moment he said, "I would think about it if I were you. These hackers are diabolical. If they want, they can also spy on you. Do you know about the webcam? The small lens above the screen? This. In theory, they can watch you from here if you leave your computer on."

I tried to remember if the computer was on the night I'd brought Daniele to the bedroom.

"So do you want me to come and take a look at the computer or not?"

But I had already turned and was on my way out of the store.

On the way back I drove very slowly. For a long stretch I was behind a truck carrying straw; the stalks broke loose and fluttered through the air. A line of cars formed behind me, but I didn't try to pass the truck. I wanted to savor the nostalgia that sight gave me as long as possible.

Once I got home I called Turin. It was my father who answered.

"I just wanted to know how you are," I said.

"Hold on."

I heard him move into another room, close a door.

"And I wanted to thank you for the money," I added.

Maybe that was a mistake. Those simple words, overheard by someone, could trigger a disastrous chain reaction. What if it hadn't been my father who'd arranged the bank transfer? I was groping my way through a plan I didn't know. I had only instinct to rely on, instinct and blind trust in Bern.

My father cleared his throat. "I keep thinking about what you wrote to me."

What he added after that, in a rushed, timid voice, caught me so off guard that I'm not sure I responded before hanging up. He spoke the most natural words for a father and a daughter, the most unnatural of all for the two of us: "You know I love you too."

For several minutes afterward, I stood there staring at the scraps of paper strewn all over the floor, as if that jumbled chaos also contained a secret message to be deciphered.

I opened a bottle of Primitivo, took it outside. The air was warm and fragrant, and I could make out the scent of peppercorns drying on the branches. The hibiscus that I had planted the year before had climbed almost halfway up the wall. Each detail produced a vivid, almost painful impression in me.

I read the book that Bern had sent me. I let the lines scroll before my eyes one at a time, until the last one. The copy had not passed through his hands, I knew that, it had come to the masseria from the shelf of a warehouse, yet I held it to my nose and inhaled its smell.

When I stood up it was dark. The screensaver's jigsaw puzzle arranged and rearranged itself in the dim light, like a breath. I moved the mouse and the monitor returned to how I had left it a few hours earlier. The translucent eye of the webcam seemed to be off. Bern was out there somewhere, he couldn't tell me, but he had found a way to let me know, the only way the bugs could not detect. Maybe he was watching me at that moment.

I dropped my jacket on the floor, I took off my sweatshirt. To slip off my T-shirt, I turned my back, I did it sensually, almost ironically, as if I were playing around in front of a mirror. Then I started swaying, just a hint of movement, not really dancing, even though I tried to hear a song in my mind.

I looked straight into the cold eye of the computer as I unzipped my jeans, as I stood there in my underwear and unhooked my bra and stepped out of my panties, certain that Bern could see me at that point, the computer's lens like his close-set dark eyes. I moved again, I tried to do it the way he would have liked, what I had done for him sometimes. For a moment his hands were on me.

EACH DAY I waited for a new delivery. And each day, when it did not come, I went into my Amazon account to check if anything had changed. But there was no further activity.

Then, one morning at the end of November, Daniele showed up at the masseria with another young man. I went toward him as he got out of the car.

"I told you not to come back."

"You have to come with us."

"Are you listening to me? You're on my property."

"There's no time. Get in!"

There was a commanding tone in his voice that made me obey him. He was already tilting the seat forward to let me climb in.

"You don't need anything, just get in," he said, seeing me glance uncertainly at the house, where my handbag with my wallet and keys was.

The wheels raised a cloud of dust as he made an abrupt U-turn. The other guy was typing swiftly on his phone. He didn't so much as glance at me.

"Did you hear?" Daniele asked me.

"Hear what?"

"They got Danco."

The guy who was texting on the phone said: "They're almost there."

"Shit!"

Daniele took a chance and passed dangerously before a curve. I leaned forward between the seats. Suddenly I was extremely tense. There was a lot of traffic on the highway, but we sped along, zigzagging to get past everyone.

"And Bern?" I asked, my mouth dry.

Daniele shook his head. "I don't know."

The guy on the phone ran his thumb over the screen to enlarge a picture. He showed it to Daniele, who nodded, then took a deep breath and looked at me in the rearview mirror again. "We have to get to Brindisi before they put him in a cell."

For the rest of the way they acted as if I wasn't there, talking on the phone with other activists, or between themselves, in a jargon that I mostly had a hard time understanding and that didn't really interest me. I leaned back in my seat. I prayed, silently, fervently, that nothing bad had happened to Bern. That he was safe.

In Brindisi the other guy stayed to keep an eye on the car while Daniele and I ran to the police station. Two police cars drove up soon after we arrived.

A knot of people had formed, blocking the steps of the building. I thought they were journalists, but when the agents got out of the cars followed by a man in handcuffs, each taking him by the arm, and when that man, Danco, smiled brazenly at the small crowd, they formed a barricade around him, linking elbows in a human chain.

Daniele pushed me forward, but I hung back. I looked at Danco, his satisfaction as he strutted toward his companions.

Again Daniele pushed me. "Let's go!"

"No."

When he realized I wouldn't move, he ran to join the demonstrators by himself. The two sides studied each other in silence. Danco acted as though he had nothing to do with what was going on around him.

At that instant he turned his head in my direction as if he knew exactly where he'd find me. He stared at me for a moment, then his lips widened in a smile that I thought seemed full of sadness.

Two armored trucks arrived and disgorged a squad of riot police who easily broke through the human barrier, creating a corridor through which Danco was made to pass. Then he vanished into police headquarters.

FROM THAT EVENING ON, the news broadcasts gave a daily report of Danco's silence. He said nothing: neither about where he had been for all that time, nor about the accomplices who may have been with him, nor the reason why he had suddenly decided to come back and turn himself in. His obstinacy astounded everyone, but not me.

Later on, it became clear that he was merely leading up to his big moment. We were already into the new year when he decided to tell his version. He did so with a statement that he read before the judge and the media, insisting that he be heard without interruption from beginning to end.

He was much better groomed than he'd been when I had seen him in front of police headquarters. His hair and beard were shorter, and he wore a gray suit, with an olive sprig in place of a pocket handkerchief, a touch that would draw sarcasm from many commentators.

He read with a firm, scathing tone of voice, his eyes moving from the paper to the judge, without a trace of subjection. He read to those who were present in the courtroom but also to everyone who would hear him in a recorded broadcast, conscious of our large numbers. He read his letter from prison, not as a confession or surrender, but as an all-points bulletin.

He described a plot behind the felling of the olive trees. He spoke of a member of the European Parliament, a certain De Bartolomeo, who had signed the first order to cut down the trees and soon afterward had designated a new species to replace the existing one. Genetically modified olive trees, resistant to Xylella, with a patent registered by a company in Cyprus, a company in which De Bartolomeo's wife, coincidentally, was a shareholder. Millions were at stake. He explained that the owner of the Relais dei Saraceni, Nacci, had given bribes to De Bartolomeo to ensure that his olive trees would be among those chosen to be felled. Perfectly healthy olive trees. All this to satisfy the needs of an increasingly blind, ravenous, and ruthless capitalism.

From time to time he sipped water from a glass. Those pauses too appeared calculated. His lawyer sat beside him, with his arms crossed and a defiant expression. Danco explained that he had always found the use of violence repugnant, and that therefore he was dissociating himself from some of the "unpleasant" events that took place that night at the Relais dei Saraceni.

"With regard to the death of Nicola Belpanno," he concluded with the same lack of emotion, "I can only affirm that I was not the one who crushed his skull. I was there, I saw what happened, but it wasn't me. And that's all I have to say on the subject."

IN WINTER, moss grew in the cracks of the concrete, a springy, shiny cushion that disintegrated at the beginning of summer, and then came back.

I painted the outside of the house because the rain had stained it with brownish drips. The scandalous drawing that Bern and I had once painted had completely disappeared, buried under layers and layers of whitewash. I scratched with my fingernails to find a trace of it, to no avail.

The last official words pronounced on Nicola's death were those of Danco in the courtroom. The prosecution had objected that the mark of the bruises on Nicola's cheek was compatible with a shoe worn by him, but the defense had shown that those hematomas were so indistinct as to be compatible with anyone's shoe. Consequently, his deposition prevailed, a unilateral testimony that, albeit indirectly, blamed Bern. But Danco was lying, I knew it. Bern's innocence was imprinted in my flesh, like the certainty of our days together.

I put in a request to the prison in Brindisi for a private interview. Though there was much resistance, the request was granted, but when I sat in the visiting room, Danco did not show up. I tried a second time, and again he didn't come. On the third request, the prison staff told me that the detainee did not wish to receive visits.

On the phone my mother kept repeating the same words: "You're still young." At first it was a consolation, "You're still young, you can start over again," but the more the months went by, the more ominous the message seemed. "You're still young, but not for long, thirty-one, thirty-two now, and you have to start all over again." But start what over again?

Moneywise, at least, things were better. A couple of guys from Noci, two dreamers, hired me as a consultant. They wanted to start a permaculture project. I didn't know if they were aware of my connection with the policeman's murder, they probably were, but still, it hadn't been talked about for a long time.

A neighbor asked to rent the greenhouse. He paid little but regularly. Plus, there was the money that my father now deposited for me every month. I learned that he now spoke of me as a skilled agronomist. For a long time I'd thought he was the only element missing in my life. During the lengthy period in which he'd refused to speak to me, I kept telling myself that if only things were put right between us, my life would be perfect. A notion that now seemed foolish.

NEARLY TWO YEARS: from the time Danco surrendered to the police, to when Mediterranea Travel, a travel agency in Francavilla Fontana, phoned to tell me that my plane ticket had been issued and that the flight was confirmed for the following day.

"You have the wrong number," I replied.

The woman on the other end took a moment to consult something, then asked: "Am I speaking with Mrs. Gasparro? Teresa Gasparro, born in Turin on June 6, 1980?"

"Yes, that's me."

"Then we spoke yesterday. You really don't remember? You told me to book the flight urgently."

A rush of adrenaline charged through my arms and legs.

"But of course. I wasn't thinking, I'm sorry. Would you please tell me the flight times again?"

"Departure from Brindisi at 20:10. You have a two-hour stopover at Malpensa. The Icelandair flight is at 23:40. It arrives in Reykjavík at 1:55."

When I'd heard the phone ring, I'd been planting the pots of strawberries. I had rich, dark soil wedged under my fingernails.

"I envy you a little," the woman on the phone said. "I was there two years ago and it was the best trip of my life. Don't miss the glacier that ends up in the sea. You can take a boat ride among the icebergs. Three days isn't enough time, but make sure you don't miss that."

I asked her if I could pick up the ticket at the agency. She said it was electronic, she had already sent the reservation to my email address. If I preferred, she could also handle the boarding passes. She asked me to confirm that I would be traveling with only a carry-on.

I don't remember how we ended the conversation, maybe I just hung up. Soon afterward I was studying the boarding pass on the computer monitor. I read the terms and conditions, written in small print, in their entirety, as if some crucial clue were lurking there too. But there was nothing more, just a seat number and an ad for a hotel that promised a discount on admission to the Blue Lagoon, with the photo of a man and a woman wrapped in towels, gazing at the horizon through the sulfurous vapors.

I should measure the sides of the carry-on bag, check the maximum and minimum temperatures in Reykjavík, pack my things, maybe do something with my hair, which in recent months I'd gotten into the habit of cutting myself with the kitchen scissors. Instead, I went outside and sat under the pergola.

The tablecloth with the world map was so faded that the top layer of plastic was flaking off. I touched the jagged, pale pink spot that was Iceland. A piece of continent adrift.

7.

The impact of the wheels touching down on the airstrip woke me. For a few seconds I could hardly move my neck. I'd been determined to stay awake through the entire trip, to register every detail before the moment I would see Bern again, but the late hour and the slight lack of oxygen in the pressurized cabin had gotten the better of me. On the stairway leading off the plane I was surprised by the very dry, icy wind. It was the dead of night, yet the sky was still bright, a radiant yellow line low on the horizon. I should have known, and instead I had imagined landing at Reykjavík in the dark.

Behind the security barrier a group of people were waiting for the arriving passengers. They wore mountain clothes, woolen caps that seemed quite odd to me coming from summer—a summer from which I'd been catapulted there. Still moving forward, I looked around for Bern, him or his black clothes among those garish colors. I looked for him in the first row, then in the one behind it; a few people looked at me meaningfully, in case I corresponded to the surname written on the signs they held up. I looked for Bern inside that tiny airport and instead I found Giuliana, standing apart by the window.

She raised a hand, not exactly a wave, more like saying, "Over here." Then she walked off toward the exit.

I reached her outside. The well-lit airport signs were extraordinarily brilliant; everything was so sharp and clear, as if there wasn't a single particle of dust to pollute the air.

"Is that all you have?" she asked, looking down at my jacket.

"It's warm enough."

It wasn't, in fact. Giuliana said shortly: "I have something in the car I can give you."

As we crossed diagonally through the parking lot, me trudging just behind her, I had so many unanswered questions, so many questions I could have asked her. The most important one being: Him, where is he? Instead we remained silent as we walked to the car, where Giuliana grabbed my bag and put it in the trunk, touching my fingers for the first time, without meaning to, I think. She pulled a windbreaker out of another bag and practically threw it at me.

We drove the first few miles through a surreal plain, the headlights revealing fluorescent lichens and small pools of a liquid that looked like milk. Giuliana said we would spend the rest of the night in a nearby village, Grindavík. It would make our trip a little longer, but not by much. It was the only accommodation she'd been able to find.

"The place we're going to is too far to start out now. We'll leave early tomorrow."

Then she asked me if I had changed my currency at the airport.

"I didn't have time."

"Usually the guesthouses accept euros," she replied, annoyed, "but they'll give you a shitty exchange rate."

Maybe it was just the haircut, a mannish razing with a very short, upright fringe that emphasized the angular shape of her skull. But studying her from the passenger seat, I was sure that something about her body had changed too. She had shriveled up. I imagined a worrisome gauntness under the red ski jacket, an extension of the thin, nervous fingers that gripped the steering wheel.

We entered a cluster of houses, all similar, so perfect with their vivid sheet-metal façades that they resembled those of a diorama. Grindavík. It gave the impression of having been built in the span of one night. Farther on, beyond an equally pristine port, you could see the solid sheen of the sea.

A very blond young man greeted us at the reception desk. Or, better yet, he didn't greet us at all, because he didn't stop watching a movie on his iPad, even while he photocopied our documents, took my money, and gave us one key card for both of us.

Giuliana said something to him, very casually, in a language that wasn't English. As we climbed the stairs, I asked her if she had learned Icelandic.

"The bare essentials," she replied.

"How long have you been here?" She fiddled with the magnetic lock that at first did not recognize the card.

"We've been here for a year and a half."

The room was microscopic, the walls wood-paneled. There was a strange smell, maybe it came from the carpet. The double bed was narrower than a normal one. As for the bathroom, it was shared. Giuliana went in before me; she didn't take long.

I brushed my teeth and washed my face and considered getting in the shower, but the plastic curtain was black with filth, crumpled up on the sopping-wet bottom. I put on my pajamas, wrong like the rest of what I'd packed, and went back to the bedroom.

Giuliana had taken off only her jacket and shoes, tossing both on the floor, and was now lying in fetal position, on the side by the window, with her back turned to me. Stock-still, as if she were already sleeping.

I hesitated for fear of waking her up, but finally I asked her: "Where are we going tomorrow?"

"To Lofthellir," she replied.

"What's that?"

"A place up north."

"Is he there?"

"Yes."

From that perspective, lying with her back to me, with her shaved head, she looked even more like a man. She did not turn around, and by then I understood that she wouldn't. I climbed onto the bed, first with one knee only, unsure whether to share that forced intimacy, then with the other as well.

"Why didn't he come?"

"He couldn't. Tomorrow you'll see why."

I don't know what came over me, why I started shaking Giuliana the way I did, repeating why didn't he come with you, I want to know why, I want to know right now; shaking her until she grabbed my arm and shoved it back with equal ferocity.

"Don't you dare touch me again," she said.

After rearranging the pillow under her head, she added: "Now get some sleep. Or stay awake, I don't care. Just keep quiet."

She went to sleep. I remained propped against the pale wood paneling. I realized that I hadn't seen a single tree on the trip from the airport to there. The smoke detector on the ceiling emitted green flashes at regular intervals. The window was covered partway down by a plastic roll-up shade, so a little light filtered in. It wasn't day, nor was it night, and I lay there waiting for I didn't know what, in that endless twilight.

WHEN I OPENED my eyes, Giuliana was putting on her boots.

"It's six o'clock," she said. "We have to go." She finished tying the laces, stood up, and opened the door. "See you downstairs."

I heard her stride down the corridor, the sleeves of her synthetic jacket rustling. For a few moments I was paralyzed, incapable of anything, then I gathered up the things I'd left scattered around. Before leaving, I took one last look around the room: the bed was only half disturbed, at some point I must have felt cold and gotten under the cover, but I didn't remember it. On Giuliana's side the blanket was drawn taut, the turned-down sheet barely wrinkled.

I went to the bathroom again and this time I clearly recognized the smell of sulfur that I hadn't been able to name a few hours earlier. It rose up from the pipes or it was the water itself.

Downstairs, the lobby was deserted. There was a coffee machine, but it was out of order. I saw Giuliana's off-road vehicle waiting by the front door, with her in the driver's seat, impatient.

"You can have these for breakfast," she said, tossing a supermarket bag in my lap.

I looked inside: a pack of triangular sandwiches and a few snacks, some familiar, others with unknown names.

"What, you don't like them?"

"Sure, they're fine."

I opened the crustless sandwiches, which were in pairs, took one out and offered it to her. Then she seemed to relax a little. After taking a bite and chewing it, she said, "All you find is crap here. After a while, you stop noticing. But up ahead we can stop for coffee. If you want."

I'm not sure I can faithfully reconstruct the conversation we had in the following hours. The words all form a solid chunk of memory, and the order gets mixed up, not only because of Giuliana's way of speaking, which often became agitated and disjointed, but also because the sleep that I had not managed to make up for during the brief night at the guesthouse was suddenly overtaking me. I'd nod off for a few minutes, and when I woke up Giuliana continued talking, or I would ask her a question. I'm sure I must have interrupted her more often than I remember, but my voice has been erased, irrelevant compared to the account of the time when she and Danco and Bern had been fugitives, and to the tortuous sequence of events that had finally brought her and Bern, just the two of them, to that island. My brain must have rearranged the information in whatever order suited it, but it doesn't matter, it really doesn't matter anymore at this point.

At each turn the landscape was less recognizable. Endless, uniform meadows, farms in the middle of nowhere, rocky expanses riddled with holes, precipitous fjords and volcanic beaches, and that one exposed road, smooth and always slightly sloping, winding interminably in front of us. The road that Giuliana preferred to speak to instead of me. At a certain point she may actually have said, "Well, I suppose you want to know," in that malicious tone I remembered. And maybe I really did reply, "Yes, I want to know."

What I'm sure of is that her arms jerked up futilely, letting go of the steering wheel, and her cheeks trembled as if she were gritting her teeth. "And what gives you the right to know?"

I looked down at my wedding ring, turned it halfway around my ring finger. Engraved on the inside was Bern's and my date: September 13, 2008.

This, I thought, this gives me the right.

"We stayed in one place for a long time," Giuliana said, still struggling with reticence, "hiding. It's a miracle we didn't kill one another in that time. Three of us shut up in a garage day and night. For months."

"In Greece," I said faintly.

"Greece? What gave you that idea?"

"That's what everyone said. That you had fled to Corfu in a rubber dinghy. Or that you had gone there via Durrës."

She shook her head. Her laugh was caustic. "I must have missed that. Well, apparently Danco's idea worked in the end."

"What idea?"

"Leaving the jeep on the coast. I was sure no one would swallow it. The life jackets on the rocks, fuck! It was so theatrical. The only thing missing was a note saying 'We left from here.'"

I recalled the images on the television news, the streets of Athens strewn with flyers. I let that backdrop fade out of my mind, then I said, "So you didn't stay in the tower."

"No, we never stayed in the tower. And we were never in Greece, we never even thought of it. We had decided from the beginning to head north."

I kept silent, an implicit encouragement to her to go on talking. Head north, where? And how far north? With whom and for how long, and to find what?

"Through one of our people we had made contact with a truck driver, a Polish guy who traveled up and down the Adriatic coast. You just had to look at him to know he couldn't have been involved with the cause, he didn't look anything like an environmentalist. He drove a semi."

"What's a semi?"

Giuliana didn't turn around this time either. "You really don't know?"

She let a moment pass before explaining it to me, just enough time for that grain of ignorance to settle between us, marking a further distance.

"It's a big rig for transporting cars. The double-decker carriers, you know? It seemed like a safe-enough solution. Someone drove us to the parking lot he would leave from."

"Who took you?"

"Daniele."

I don't know why, but at the moment I didn't register how odd it was that she would mention Daniele, as if she took it for granted that I knew him.

I felt slightly nauseated, the taste of the sandwich was sour in my mouth. I felt overcome by the urge to sleep and at the same time gripped by an uncontrollable agitation.

Giuliana went on: "I didn't know what had happened. Only that Bern and Danco had come running down through the olive-tree grove shouting, 'Let's go, move!' And then we were in Daniele's car, and Danco was constantly turning around to look through the rear window, while Bern, seated up front, didn't do so once. He kept his hands oddly on his knees, as if they weren't his. And even later in the parking lot, as we waited for Bazyli, there was something strange about his posture, a kind of stiffness. He asked me for a cigarette and I gave it to him and only then did I see why he had kept his fingers pressed to his knees the whole way: his hands were wounded, there were two patches of dark blood, already dry. I rubbed them with a handkerchief, but without water the blood wouldn't come off. So Bern spit on them. He held out his hands with a docility that wasn't like him, they were . . . limp. I asked him if I was hurting him, and he said not to worry. I cleaned the blood off of one hand and then the other; there was no wound underneath. I looked at him, but he remained blank, letting the information pass from his eyes to mine in silence. I too lit a cigarette and we waited there in the parking lot with nothing more to say. It was Danco who explained that it had been an accident: 'They attacked us.'"

Talking about the cigarette that she had smoked in the parking lot on that crucial night made Giuliana search for the pack in her jacket pocket and reach for the car's cigarette lighter. She touched the glowing circle to the tip with a precise gesture. Only after she had exhaled the first stream of smoke from her nostrils did she ask me if it would bother me at that hour.

"No, go ahead," I told her. She lowered the window a few inches, and after that she blew the smoke into the opening.

"Bazyli wasn't surprised to see us," she continued. "He didn't ask any questions. He simply repeated the name of the destination where he would drop us off. A figure had already been agreed to, two hundred euros for each person he would transport, and he demanded that we pay him the money immediately. Each of us had gone to the Relais supplied with cash, because we had no way of knowing how things would go. Bazyli showed us which cars we would travel in, one person per vehicle, because we had to lie down and not raise our heads for any reason. I got a white Citroën and saw Danco and Bern get into theirs. We didn't exchange a word, no goodbye or good luck, not even a look. The seats were covered with nylon sheeting and I lay down. Even before hearing the clanking as the semi started out, I had fallen asleep.

"I woke up because of the cold. Two or maybe three hours had passed, but the sky was light, a dense white. The inside of the car was an icebox. I huddled up, trying to wrap the nylon around me, but it didn't help. Bazyli had mumbled that the trip would take sixteen hours or so, I would never make it at that temperature. Plus I had to pee. I held out for almost an hour, but finally I couldn't take it. Bazyli hadn't mentioned any stops, hadn't left a mobile number, and anyway, Danco had seized our SIM cards so we wouldn't be tempted to use our cell phones. There was nothing else I could do but crawl between the two front seats and start sounding the horn, for ten whole minutes. Finally I realized that we were pulling over, that we were stopping, but it took quite a while before Bazyli opened the door. 'What the fuck are you doing?' he laid into me. When I explained that I needed a toilet immediately, he helped me climb down and told me to hurry up.

"Outside the bathroom I met Danco, but we pretended not to know each other. It's strange, we hadn't decided on it before, we hadn't agreed on that point, but there was a kind of instinct that guided us. I let him see that I was cold and he went into the service station to look for something that I could put on, but all they had were kids' raincoats with ridiculous designs of superheroes. He took two from the display rack. I had no more money. Near the exit I touched a package of snacks and then put it back. Again, Danco understood. He bought the snacks and a box of crackers. We went out one at a time.

"When we went around the corner to return to where the truck was parked, we saw Bazyli with two policemen. He was explaining something, gesturing. I was so scared, I couldn't move, until I felt Danco's hand clutching my arm and pulling me back. We stayed like that, practically not breathing, leaning against the wall that the police might appear around at any moment. Bern hadn't left his hiding place. By this time they might already have found him. I told Danco that we had to get out of there, climb over the guardrail and run through the fields. He said leaving Bern there by himself was out of the question. When we peeked around the corner again, there was no sign of the cops. Bazyli was waiting beside the semi.

"'What did they want?' I asked him, but he motioned for us to get back to our places as quickly as possible. He handed me an empty plastic bottle, saying, 'Next time, here.' Then he pointed to the packages of food I was holding and with an unmistakable gesture let me know that he would strangle me if I spilled any crumbs in the car. I think he meant it."

GIULIANA CRUSHED the cigarette stub in the ashtray between us. There were others in there, the little pile reeked. She must have noticed my glance, because she said: "Ten years of decomposition, I know. And here on the island cigarettes cost a mint. But this is no time to quit."

She closed the lid on the ashtray.

"Do you have any chewing gum?"

"No."

Her eyes twitched constantly, a nervous tic that I didn't remember her having before. To avoid a truck coming from the opposite direction, she swerved too far to the right, and the wheels ended up on the dirt shoulder. A pebble flew up and hit the windshield.

"And do you know how long it takes for chewing gum to decompose?"

"No."

"Five years. Alkaline batteries?"

"I don't know."

"Go on, take a guess."

"Can we drop the riddles?"

Giuliana shrugged. "We used to play that game in Freiberg. One of the many ways to pass the time."

"Freiberg?"

"At Bern's father's place. We stayed with him. He sent a friend of his to pick us up where Bazyli had left us. He had us brought to a garage of his."

"Bern hadn't seen his father since he was little," I said.

Giuliana turned for just a moment.

"Well, maybe he didn't see him, but he certainly was in touch with him. Otherwise he wouldn't have had the phone number memorized. Maybe he didn't want it known. Bern can be very private about certain things. I imagine his father is one of those things, and I certainly can't blame him."

Talking about Bern that way, letting me see that by then she knew him better than I did, gave her heartfelt pleasure. But I couldn't help asking her why.

"His father is not the next-door neighbor that everyone would like, let's put it that way. He mostly deals with reselling works of art of somewhat dubious origin."

"Stolen?"

Giuliana put her thumb to her mouth and bit her cuticle.

"I'm convinced he fences them for someone else, otherwise he'd be much wealthier. But he has a storehouse full of the stuff. Masks, vases, sculptures, and so on. All heaped together in that garage, which for some reason has a bathroom and a small refrigerator. And an internet connection. He must have spent fairly long periods of time there. Anyway, that's where he kept us. For eight months."

Before our wedding, when I had asked Bern if he wanted to track down his father to invite him to the ceremony, he had stared at me with an incredulous expression and shook his head, as if the idea were absurd. Instead, his father had always been in the same city, Freiberg, and he spoke to him on the phone. When had he done that? At times when we were not together? When he went off among the olive trees alone, as though responding to an irresistible call of the countryside?

"Our German period," Giuliana said, and chuckled. "Freiberg. Not that we got to see much of the city. Sometimes we went out, we took turns, but we had to be very careful. The German didn't want us to."

The German. The grave-robber, my father had told me. An unexpected sadness came over me. Giuliana didn't notice.

"Danco was uneasy. The business about the artworks bothered him. All that stuff should have been in a museum. By staying there, he said, we became accomplices. As if that were the real problem. But he wasn't very lucid. He'd wake up in the middle of the night, unable to breathe, and yank the sheets suddenly, uncovering all three of us. Then he'd start pacing around the room, gasping for air. Something like that had happened to him at university too, panic attacks due to the exams, but it was nothing compared to Freiberg."

"All three of you slept together?" My mind had focused on that insignificant detail.

"There was only one bed," Giuliana replied expressionlessly.

"Danco said it wasn't he who killed him."

After pronouncing those words, my cheeks began to tingle, then the tingling became a burning sensation and spread to my neck and arms.

"Have you seen his lawyer?" Giuliana asked. "Hired by his father. Viglione Senior couldn't wait to make himself useful. Danco always knew he had his back covered. But I imagine that's how it is for all of us: ultimately we go back to where we started. Which is a rather serious problem for me."

She burst into caustic laughter. I remembered her stories, from when she and Corinne would vie to see which of them had had the worst childhood. None of that interested me now.

"He was lying, wasn't he?" I asked.

Giuliana's fingers flicked up for a moment, then settled again on the steering wheel.

"Who can say?"

"You can. You were there. And afterward you were with them the whole time."

"Sorry. I can't help you with that. I understand that it may seem important to you, but to me it isn't."

I sensed a tension in her body now, as if she were preparing for a fight. Or was I the one spoiling for one?

"It's not important? You mean to say you never asked him what really happened?"

Giuliana shook her head. She continued looking at the road. I may have leaned toward her.

"You lived in a garage for eight months, you slept in the same bed all that time, and you never talked about that night?"

"What's done was done. Would talking about it change anything? We were all there at the Relais. We three and thirty others. It could have happened to anyone."

"You're joking!"

"You're getting a little too worked up, Teresa."

"A man is dead! A man I knew!"

"Yeah, Bern told us that. You had an affair, you and that cop."

"Did you ask Danco how things went at the Relais or not? Did you ask Bern or not?"

Giuliana absently brushed her hair back, what was left of it. She seemed surprised that it wasn't as long as it used to be.

"Let's stop here," she said, taking an exit off to the side. "We have to get gas. I hope you have some cash left."

INSIDE THE SERVICE STATION we split up. There wasn't a real bar, just a corner with tall coffee dispensers and a stack of paper cups next to them. The price, indicated on a sign, was to be paid at the cashier. If you drank the coffee and left without paying, no one would have noticed, but that probably didn't happen on the island.

I wandered among the shelves awhile, the same souvenirs I would see again and again in the days that followed, but that at that moment were new: seal-shaped stuffed toys, thick wool sweaters with traditional designs, hats with Viking horns, and miniature trolls.

A large map of Iceland, slightly yellowed, hung on one wall. Framed pictures highlighted the tourist attractions: geysers, volcanoes, waterfalls, all names impossible to pronounce. In one photograph I thought I recognized the icebergs in the sea that the woman at the travel agency had told me about. For some reason I was sorry to miss them.

"We're almost at Blönduós," Giuliana said. She was holding two containers of coffee and handed me one. She pointed her finger at the map. "We're traveling this road. It forms a ring around the island. We have to get here."

A lake, located almost in the center, to the north.

"Mývatn," I read.

Giuliana corrected my pronunciation, then explained something about the construction of words in Icelandic. All of a sudden I saw the absurdity of being in that shop, in that remote place on earth, with a person who by then was in all respects a stranger to me, surrounded by refrigerator magnets commemorating the volcanic eruption that a couple of years earlier had covered half of Europe with ashes. Yet being there with Giuliana, traveling to a destination that I couldn't even pronounce correctly, was one of the first vital experiences I'd had in a long time.

"Why there?" I asked.

"We were looking for a place that hadn't been corrupted by man. Something intact."

"And did you find it?"

Giuliana turned away abruptly, her back to the map and me.

"He found it, yes. Let's go."

FOR A WHILE we rode in silence. I stared at a formation of clouds to my right, a high, puffy mass, like a nuclear explosion, motionless in the sky. Even the clouds seemed different here. No matter how far we drove, the cumulus did not move with respect to us: impossible to reach, impossible to go around, impossible to avoid. Then Giuliana said: "It was a delicate balance. You have to try to understand that. None of us had ever been in such a situation. No one had ever even imagined it."

She took a deep breath. She accidentally turned the windshield wipers on, and they scraped against the dry glass, screeching. For a moment she seemed thrown into confusion.

"A week after we arrived, the German came to see us. I would have recognized him even if he hadn't opened his mouth. He stood in front of Bern, and the resemblance was astonishing, with the exception of their coloring, because the German's hair was mostly white and he had light eyes. He spread his arms, Bern went to him and let himself go completely.

I don't know why that gesture moved me, we were all still upset, a week had gone by and we'd never left the place, shut in there with no news, left in suspense, with only someone who brought us food and didn't say a word. And now here was Bern's father, and Bern letting him hug him like a child.

"The German shook hands with me and Danco. He asked us if we were getting bored. It had never even occurred to us to be bored. Then he asked us if we had used the computer, but that hadn't occurred to us either. So he sat down at the desk and explained that we could surf the net securely. He had a firewall like that of the Pentagon, plus an untraceable IP address. He did not specify that all those exceptional precautions were due to the illegal trafficking in works of art; we realized it ourselves, or at least I did, and I think Danco did too, whereas Bern may have just been bewildered. The German sat at the computer, with me behind him, Bern behind me and Danco in back; he wanted to keep his distance even though he was intrigued. After days of stagnation, that was our first distraction.

"The German asked if any of us was familiar with Tor. I was, because at university many of us used that software, mostly to buy weed; those were the years when playing hacker guaranteed you a certain fame.

"'So sit here in my place,' he said. 'I'm afraid you kids will need a transformation, unless you want to stay in here forever. We can't change your faces, but we can at least find you a new identity.' He got up and I sat down at the computer. The procedure was simple enough. We just had to take photographs, upload them, and in a few weeks we would receive three brand-new passports. We could choose whatever nationality we preferred, but the German advised us that unless we spoke another language perfectly, it was better to stay with Italian. He had brought us a camera. The documents would be sent to a postal box where he received all his shipments. He conveyed an incredible calmness as he spoke, and seemed almost amused. He stayed with us a while longer, telling us about the system used to authenticate the works in the garage and sell them online. A complicated system in which he obviously took great pride. Then he promised to come back and see us as soon as possible. Before leaving, he ruffled Bern's hair, as any father would do to a son.

"Bern suggested we all choose the same surname, as siblings. We discussed that possibility, but it didn't seem advantageous to me; it would have added an element of risk. Then we talked about where we would go once we got the new documents. Not somewhere in Europe, that was obvious. Bern and I proposed destinations, explored them on Google Earth, and every night we convinced ourselves that we had found the ideal place: Cuba, Ecuador, Laos, Singapore. In the morning we questioned everything again. By then Danco refused to participate. He kept saying that the forged passports were dirty. Which criminal organizations were we implicitly asking for help? He adhered to his rigid, unassailable principles of morality. He couldn't see what Bern and I were all too well aware of: that by then we had overstepped that ethic. From our standpoint, the morality that Danco defended was simply inadequate."

GIULIANA REACHED OVER to the backseat, where her bag was. She fumbled inside, but couldn't find what she was looking for, so she grabbed it and put it on her lap.

"Here," she said, handing me a passport.

The face, under the shimmering reflections of the laminated paper, was hers, her hair already shaved. Next to it, her new name: Caterina Barresi.

"When we get there, please call me that," she said somewhat gravely.

"What name did he choose?"

"Tomat. It's Friulian, quite recognizable, and you can say what you want about Bern's accent, except that it sounds Friulian."

We were at the end of a fjord. In front of a rock cliff were two twin houses with sloping roofs, a distance away.

"Danco and Bern were barely speaking to each other. Deep down, Danco held Bern responsible for what had happened. Actually, they were on bad terms even before the Relais. Danco did not accept the idea of using weapons. They were contrary to everything he had always believed in, he said, and that was true, I knew it was. But they were also contrary to everything I had believed in, or Bern for that matter, or all the others who were at the encampment with us. Could we help it if they had become necessary? Sometimes you have to go beyond what you think is right in order to achieve a higher goal. The new order must come about through disorder, that's what Bern made us understand, but not Danco, no, he refused."

I thought about the day Danco had showed up at the masseria with Bern's list of things to take, about the determination in his eyes, a rancorous determination that I hadn't understood.

"But one day, at the encampment in Oria, Bern took him on a walk through the felled olive trees and convinced him."

Giuliana lowered the window, put an arm out to resist the cold air, then leaned over to offer her face to the wind as well.

"Or so it seemed at least," she added in a bitter tone. "Can you drive a little?"

I really didn't want to. The drowsiness was still there, mixed with the acidity of the sandwich and the horrible coffee I'd just drunk. And I knew I wouldn't dare maintain Giuliana's speed on that road, where every curve seemed about to eject us.

"I just need half an hour. To close my eyes," she insisted. We traded places. Before getting back in the car, Giuliana bent down and grabbed her ankles, and stayed like that for about twenty seconds, her muscles taut under her jeans.

For the first few miles she kept her eyes closed and her head erect like a statue, but she wasn't sleeping, I knew it. When she opened them again, she said, "I miss the olive trees. I miss just about everything. Especially the warmth. Summer didn't even last a month here. It's because of global warming. The ice that melted in Greenland has cooled the Gulf Stream. While the rest of the world is melting in the sun, we end up freezing, even in August."

"I was at the encampment," I said then, maybe to console her, or just the opposite, because I wanted to intensify her nostalgia.

"I know."

"You know?"

"Daniele told me."

"You talked to Daniele?"

Giuliana glanced at me. "I talk to him almost every day. Why would he have gone to look for you otherwise?"

Her mood changed again. Suddenly more conciliatory, she added: "We resumed communication a couple of months after arriving in Freiberg. It wasn't the least bit easy. If my computer skills were rusty, Daniele's were pitiful. And there was the problem of sending him the first message without leaving a trace. I got the idea of using Amazon. Daniele was under house arrest, so it was entirely plausible that he would order things via the internet. So I had him buy an electric toothbrush, since it was an item that we had joked about together some time before. He'd told me that his mother made him bring one with him everywhere, even to the encampment. There he was, out in the countryside, wandering around with that buzzing gadget. It took me a while to get access to his computer and his credit card. But when he received the electric toothbrush he understood. I sent him the instructions in a series of emails that anyone would have mistaken for spam. And within a few days we had a protected network available so we could write to each other directly."

She propped her foot on the dashboard and slid lower in her seat.

"I don't know why I'm even telling you these things. You could go back to Italy and report everything to the police."

"That's what Bern did with me too," I said. "He sent me a plant product and a book."

"Bern and I sent you those things," Giuliana clarified, giving me an ironic look. "Or, rather, as far as the fertilizer is concerned, me, Bern, and Danco. On his own, Bern wouldn't even have been able to boot up the computer."

"But why not let Daniele tell me where you were? If you were already in contact with him."

"Damn, why didn't I think of that!" she said, and started laughing.

"Why, then?"

"It was he, Daniele, who didn't want to. He'd studied you for a while and in the end he decided you couldn't be trusted."

He'd studied me. He'd come to bed with me.

"Were you able to see me?" I asked, feeling more and more tense.

"When your monitor was on, yeah. If you ask me, you have a pair of my panties."

She laughed again, a spiteful laugh, a little strained. I slowed down and stopped the SUV in a rock-strewn pull-out.

"What are you doing?"

I got out and started walking through the tract of heather. The island was barren, you couldn't see anything in the direction in which I was going, absolute emptiness, I could have walked on endlessly without encountering a single obstacle. I heard the car door slam.

Giuliana shouted, "Hey, come back here! I'm sorry, I didn't mean to upset you. Come on back!"

But I didn't stop. The topsoil among the plants was dark, almost black. Giuliana must have started running. Soon she was in front of me, blocking my way.

"We still have a long way to go. If we lose time we'll have to wait until tomorrow. And tomorrow may be too late."

"Too late for what?"

I kept going, so she was forced to walk backward.

"You'll see. Now let's go."

"Where is Bern? I won't get back in the car until you tell me where he is."

"I told you, you'll see."

I screamed at her: "Where the fuck is he?"

"He's in a cave."

"A cave?"

"He probably won't be able to hold out for much longer."

I stopped walking then. Giuliana stopped as well. The wind lashed us from the side, not in gusts like Speziale's tramontana, this was a constant force.

I wasn't all that surprised, actually. Bern inside a cave: I could imagine it. In all those years I had grown used to his bizarre quirks. He'd lived in a ruined tower, in a house without electricity, up in a tree. All I asked was "For how long?"

"Almost a week."

"And he can't come out?"

"No. He's stuck in there."

That stubborn, angry wind, the quivering tufts of heather clinging to the stones. Giuliana reached out to grab the edge of my jacket.

I let her lead me back to the car. She got back behind the wheel. I huddled at the far end of the seat, as far away from her as possible, but we didn't drive for long. She parked in front of a building larger than the others, in a prominent position on a hill.

"We can get something decent to eat here," she said. "I think we could use it."

INSIDE, a fireplace was lit. Heads of cloth animals were mounted on the walls, as though that alpine decor were a parody, since no one would have dreamed of stuffing real animals. We sat down in a corner, my back to a window. I felt a weariness spread through my limbs. When they brought us the menus, I didn't have the energy to look through mine. It seemed like too purposeful an act, too normal. An unspoken question, unutterable, held me frozen: if he can't come out, what will happen to him?

Giuliana ordered for me. They seemed to know her. But they probably knew Caterina, not Giuliana. The girl serving the tables, her face and gestures so open, brought us two white cream soups. Dark pieces floated on the surface.

"It's mushroom," Giuliana said. "I hope you like it."

I must have been very pale, or maybe there was something more serious than the pallor about me, something troubling, because such concern wasn't like her. I don't remember if it was she who guided my hand to the spoon or if I did it instinctively, but I ate the soup, spoonful after spoonful, the morsels of mushroom chewy under my teeth, tasteless, like bits of styrofoam.

I felt a little better afterward, but I didn't taste even a bite of the salmon that came next. Just seeing it made me suddenly sick. I rushed to the bathroom and threw up.

I stood in front of the mirror for a long time, looking at a face I didn't recognize, cheeks reddened by the restaurant's heat or by the cold outside, or maybe by the shock. When I returned to Giuliana, the table had been cleared. She asked me if I felt better; I didn't answer.

She signaled the girl, who appeared a few minutes later with the bill. As on all the other occasions, she waited for me to take out my wallet and pay for both our meals. When I was about to pick up the Icelandic crowns left as change in the small saucer, she stopped me with a wave of her hand.

"Leave it as a tip."

IT HAD STARTED to rain, a light, very fine drizzle. Looking at my sleeve, I realized that it wasn't really rain, but sleet. At the end of August. I remembered when Bern, in the parking lot in Kiev, had walked to a rotunda covered with icy snow and placed his hand on it, the wonder in his eyes.

"Why did you send me those things?" I asked. "The pesticide, the book. Why, if you didn't trust me?"

"It was Bern who insisted on it. You looked so dejected, he was worried about you. And about the tree. On those forums where you asked how to treat it, the answers you got were just a load of crap. Obviously it was Danco who found the pesticide. It was one of the rare times when he went to the keyboard and interacted with us. By then he barely spoke to us. At night he had terrible nightmares, or he didn't sleep at all. I had asked the German to bring me some sleeping pills and I would sometimes crumble them into his food. I'm a little ashamed when I think of it, but I did it for him. I was really afraid that he was losing his mind."

"So Daniele knew where you were," I said. I couldn't let go of that thought.

"We desperately needed to do something. By then we had our passports, our new immaculate identities. Daniele sent us photographs of the olive-tree reserve at the Relais dei Saraceni, or what used to be the reserve. Craters instead of trees. In the first weeks, at least people talked about us, hunted for us, we felt movement around us, but once autumn came the monotony became unbearable. 'We're sitting here twiddling our thumbs while they destroy everything,' Bern kept saying."

She sighed, as if she'd told that story dozens of times already and repeating it was getting to her.

"One day I hacked into Nacci's computer. Exactly as I'd done with Daniele and with you. I had gotten the hang of it and his password was so obvious that I guessed it on the fifth or sixth attempt. I found all sorts of filth there. And in particular I found his correspondence with that European Parliament member, De Bartolomeo. Proof that we were right from the beginning, about the golf course and everything else. If someone had only listened to us, well, none of what happened would have happened."

Suddenly, her willingness to tell all made me furious.

"What are you really sorry for? Are you sorry that Nicola was killed? Are you sorry about the olive trees? Or are you only sorry for yourself?"

For the first time Giuliana looked at me with a hint of uncertainty.

"The olive trees were the most important thing," she murmured.

"The olive trees? Do you really believe that olive trees are more important than a human life?"

"At the time I thought so. We all thought that, I think. Maybe we were wrong."

Yes, you were wrong. You sure as hell were wrong!

But I didn't say that. What I said, with the same accusatory tone, was: "You had explosives."

She shrugged, as if it no longer mattered. She remained silent for a few minutes before continuing: "The investigation into Nacci's affairs seemed to awaken Danco. He had already decided to turn himself in, but none of us could have imagined it."

Giuliana's arm was propped against the window and her head was resting on her arm.

"One morning Bern and I woke up and Danco wasn't there. None of us was supposed to go out without having first discussed it with the others, even the German had recommended that, and especially not at that hour, when the streets were full of people on their way to work. We sat and waited for him for a couple of hours, then Bern couldn't stand it anymore and went out looking for him. When he came back he looked terribly weary, forlorn. He'd understood.

"Something changed in him, afterward. I don't know why exactly, but I think it was seeing Danco on television, handcuffed, letting himself be led into the police station. 'Doesn't he seem free?' he asked me. 'Free?' I repeated, pretending not to understand, but it was true. Danco in handcuffs looked free, much more so than we two were, stuck in that garage. But there was no time to think about how we felt. We had to leave immediately. Danco might already have revealed our hideout to the police. We packed our bags. Bern preferred that we didn't notify the German. He wrote a goodbye note, stood staring at it for a long time, then crumpled it in his fist. He couldn't even say goodbye to his father the way he wanted to."

GIULIANA HAD TEARS in her eyes, overcome by tenderness. And seeing her so moved, lost in that memory of Bern and his father, I understood. Not as though I were realizing something that had earlier been wholly mysterious or inconceivable, but the way you catch a puff of dust fluttering through the air after having followed it for a long time with your eyes. I understood what I already knew and had never wanted to admit. I'd known it from the moment I had looked for my husband in the small crowd at the airport arrivals hall and instead of him I'd found her.

I said, "Why did you cut your hair like that?"

She made that gesture again, which I'd seen her repeat numerous times since the night before, her fingers searching for the long hair that no longer existed.

"I don't know."

"To be less recognizable?"

"No," she said, but corrected herself immediately. "Maybe. I thought . . . I liked it better this way."

"It's Bern who likes it like that, right?"

Yes, I'd known it long before I landed on that cold, remote island. The hostility with which Giuliana had welcomed me to the masseria, which had never really abated, the way she stared at Bern, the habit of putting her hands on his back at the end of the day, massaging his neck and shoulders, and him closing his eyes; a friendly gesture, nothing more than that, I told myself, yet each time I had to find something to do, so I wouldn't see them, wouldn't see the expression of surrender that came over his face.

"You slept together."

And since Giuliana still didn't speak, conceding only mute assent, it was I who said: "It happened even before. Before I arrived at the masseria."

"What difference does it make now?"

She looked for the pack of cigarettes, took one out, lit it. Her fingers were shaking.

"And while I was there?"

"Stop being paranoid."

I grabbed her arm. I squeezed as hard as I could. I didn't mean to hurt her, I just didn't want to let her get away, as if her body were bound to the truth that she refused to tell me. Giuliana stiffened her muscle but didn't try to break free.

"I have a right to know," I said softly.

"Only twice. At the beginning."

I let go of her arm, sat back.

"And Danco?"

Giuliana shrugged, a movement that could mean she didn't give a damn about Danco anymore. Or it could mean that Danco had known about it. That the whole story, his abrupt rift with Bern, and maybe even his arrest were linked to that awareness.

All of a sudden I felt a familiar, distancing sensation wash over me: of objects moving away and getting smaller, except that this time it wasn't them moving away, but rather me: I was receding at a frenzied speed, farther and farther back, in a tunnel that had opened up inside my head.

"Stop!" I ordered Giuliana, but she continued driving and I didn't have time to say it a second time before the first acidic surge rose from my stomach and filled my mouth; I held it in with my hands as Giuliana jammed on the brakes. I opened the door and vomited the rest of the soup, all those venomous mushrooms.

She handed me a tissue, and when I didn't take it, she laid it on my knee. I used it to wipe my mouth.

Then I leaned back in the seat again, my eyes closed, the beating of my heart slowly returning to normal. With a nod I told her she could continue on.

WE REACHED the lake a couple of hours later. The sky had opened up, you could feel summer now. A dense vapor issued from a fissure on a barren mountainside. There was the smell of sulfur there too, stronger than in the guesthouse.

We skirted the lakeshore for a while; the surface of the water glittered, and tiny green islands sprouted here and there. This was a more recognizable setting, more reassuring than the miles of uninhabited, alien nature we'd driven through in the preceding hours.

Giuliana turned into a slightly sloping parking area. She switched off the engine.

"The bathrooms are in there."

I felt weak, dazed. I asked if we were stopping there.

"We have to change vehicles. With this one we can't get to the cave."

The new off-roader had gigantic, overly inflated tires, as if someone had blown them up disproportionately as a joke. It belonged to a guided tour agency, a name with the word "Adventure" in it, or "Outdoor." On one side it had the image of a group of people rafting, with splashes of foam surrounding their smiling faces.

Giuliana introduced me to our guide, Jónas. He was no more than twenty-five, and was in short sleeves despite the temperature, a waterproof jacket knotted around his waist. They spoke in a rapid, clipped English that I wasn't able to understand. Then Jónas asked me with the utmost cordiality if I had gloves and if the shoes on my feet were the only ones I had. In both cases, Giuliana answered for me: I would use her gear. Jónas helped me climb onto the jeep's high footboard, while she watched from below, and a moment later we were on our way.

We took the road that circled the lake, back the way we'd come. We continued on for another half hour or so past the point where we had turned in, before Jónas turned right onto an unmarked dirt road. Giuliana and I were sitting in different rows. There were a dozen or so seats in the jeep, all empty except ours.

I was observing the landscape, beginning to get used to the vastness. I found myself imagining what Bern had felt when he saw that terrain for the first time, the wonder that must have riveted him, because in him the sense of wonder was always boundless.

"We were looking for a place that had not been corrupted by man. Something intact," Giuliana had said earlier.

I wanted to ask her to explain it to me better, but I couldn't stand the idea of hearing her talk about Bern again, not then.

After a few miles the road became rougher. The mule track at the beginning had narrowed to double dirt ruts, barely visible, which had probably been made by the abnormal wheels of the same jeep we were traveling in. Grass grew in between. It reminded me of the dirt track at the masseria, but a treacherous, forsaken version, as it might have looked after a flood. There were depressions and dips and protruding boulders; the jeep swayed on its suspensions as if about to overturn.

In the rearview mirror, Jónas signaled to me to hold on to the rubber strap hanging from the ceiling, and I grabbed it a moment before an even deeper hole bounced me out of my seat.

A little farther on he stopped, got out of the vehicle, and bent over to examine one of the tires. I saw him go around and open the rear door. He returned to the wheel with a toolbox.

"Did we get a flat?" I asked Giuliana. Instinctively, I turned to look at her, and that simple gesture seemed to sanction a truce. I immediately regretted it.

But she barely looked at me: "He has to decrease the pressure to improve the traction. The road gets worse from here on out."

When Jónas had completed the operation on all the tires, we set out again. It seemed impossible to me that the condition of the track could get any worse, but I was wrong. In the hour that followed, I had to hold on tight to the strap with one hand and grip the base of the seat with the other.

The shaking continued even after the bumpy road ended abruptly and we kept driving on a soft carpet of dark sand, at the base of a volcano. That tremor came from within now, from the fear of finding myself with Bern after all that time. The sky was even stranger than it was elsewhere, a dull blue, scored with white streaks that crossed in all directions.

Jónas repeated the operation with the tires, in reverse. I stared at the low trees, similar to rhododendrons, that grew at the foot of the volcano. Then I spotted a trailer in the distance, the only trace of humanity in the midst of all that emptiness.

INSIDE THE TRAILER, climbing boots were arranged by size on wooden shelves. On the opposite side, tossed chaotically in a box, were mud-splattered safety helmets.

"We'll change our shoes," Giuliana said.

"I can keep these on."

Given the situation, accepting a favor from her was unthinkable. But the severity with which she answered made me bend down and untie my Adidas. I put on her hiking boots.

"Cross the laces at the top. Pull them tighter," she ordered me with the same commanding tone as before.

Then Jónas gave me a pair of snow boots, a helmet, and heavy wool socks that smelled of sweat. He explained to me that I would put them on before entering the cave; we had to walk half an hour to get there. He pointed in the direction we would set out.

"Lava camp," he said, turning to the expanse of wide, flat boulders that stretched before us. Small canyons ran through it like veins. The cave was somewhere in the middle of it—Bern was somewhere in the middle of it.

The walk was longer than predicted. Maybe I was slower than Jónas expected, or maybe we took a circuitous route, because they too seemed to be proceeding instinctively, with a precise destination point in mind but no fixed course through the rocks.

I was exhausted, yet tension kept me going. I stepped on a rock and my foot twisted. Giuliana reacted quickly to support me from behind and keep me from stumbling, but I had to stop for a few minutes. Jónas crouched down in front of me and made me place my leg over his knee while he unlaced the boot and moved my foot cautiously from side to side. He asked if I could continue, telling me I needed full mobility to enter the cave. My ankle hurt, but I said yes, then I did my best not to show that I was limping.

At last, at the entrance to the cave we found two other young men. They had set up a tent and were sitting at a camp table, with two thermoses in front of them. Introductions were made hastily. They exchanged a few words with Giuliana about the delay, maybe we shouldn't go, they said, better to wait until tomorrow. Giuliana insisted. They agreed, but said we would have to be out within an hour.

As they debated, I approached the edge of the crater, which was about ten yards wide, yet invisible until you reached it. At the bottom a gleaming mantle of shimmering moss covered a heap of stones, probably the remains of the collapse that had created the opening. On one side ran an iron ladder, with only a rope to act as a handrail. I took a step forward to get a better look, but I got dizzy and had to back away.

I didn't listen to much of what Jónas advised. My desire to go down there was as intense as my desire to get away from there as soon as possible, to go home. I understood that the inside of the cave was covered with ice, that the soles of the snow boots were studded with cleats to provide traction, but that I would have to be careful. Jónas asked me if I suffered from claustrophobia; he had to repeat the English term, twice.

Then he and I went down, him in front. Giuliana did not follow us, but remained at the entrance with the other two guys. I went back up the few steps. "You're not coming?"

She had her arms folded, and her eyes had dark circles under them, or maybe it was just the light.

"He wants to talk to you," she said. "That's what he asked me, so go."

Then she turned, and I saw how much it had cost her to say those words, how much it had cost her to meet me at the airport in Reykjavík and share a bed with me and then be in the same car together for ten hours, all to bring me to the man for whom we'd silently been rivals for years. I felt sorry for her.

At the bottom of the ladder the light was poor, but you could see the metal gate that marked the entrance to the cave. We stopped a few steps from the ice. Jónas told me to put on the woolen socks, the cleated boots, and the helmet with the front-facing light on top, which he switched on for me. He also had an additional sweater. I was already warm the way I was, but he forced me to put it on, since the temperature inside was close to zero. Soon enough I'd see what he meant.

The entry was the hardest part, but I didn't yet know that. You had to clamber up a slippery boulder and crawl on your belly through a slit about half a yard wide. Jónas went ahead of me, showed me how to do it, but it took me five attempts. Then I had to make my way through a tunnel, stooped over. I couldn't breathe, and felt my heartbeat accelerate wildly. Maybe what I'd told him wasn't true, maybe I did suffer from claustrophobia. I remembered the night when Bern had taken me to the tower, the dark steps and the panic that had made me beg to leave there as quickly as possible.

The ice was thick and the beam of light revealed forms trapped inside it, colorful stones made brilliant by that crystalline layer.

When the tunnel sloped downhill, Jónas told me to let myself slide, holding on to the rope. He would help me with the landing. For a while my arms refused to loosen their grip on the rope, but I heard Jónas encouraging me, his voice terribly distant, and I let myself go.

Finally we found ourselves in a chamber, a large cavern with ice as its floor and dark rocks overhead. Jónas warned me to be careful not to bump into the stalagmites that were scattered everywhere; some were a few inches high, others came up to my forehead. He said it had taken hundreds of years for them to form, but that just brushing them with my foot would be enough to shatter them. I had to step exactly where he stepped.

At first, I proceeded very slowly, until I became familiar enough with the slippery floor. We crossed the chamber and entered another through an opening in the rock. I looked around to gauge how big it was. It was smaller than the first one and there was no visible way out this time. It looked like the terminus of the cave.

Jónas raised an arm and pointed to something ahead of us, up above, and only then was I able to make out a very narrow, horizontal cleft.

"He's in there."

He cupped his hands around his mouth and called Bern's name. It produced an echo that seemed never-ending.

The silence had not yet been fully restored when Bern answered: "Yes."

Then I could no longer contain myself; a gush of emotion flooded my chest and tears streamed from my eyes. Later, much later, remembering that moment, I would think about how those tears spilling onto the ground had joined the perennial layer of ice, but not at that moment. At that moment there was only Bern, beyond a rock whose thickness I could not imagine.

Jónas helped me climb a couple of yards closer to the cleft. He pointed to a ledge where I could sit. Going any farther up would have been impossible for me, but from there Bern was able to hear me, I just had to speak loudly. Jónas would stay at the base of the chamber; he couldn't risk leaving me alone.

"Bern," I said.

There was no answer. Jónas told me to raise my voice. I repeated the name, nearly shouting.

"There you are," he answered then.

I had the impression that he was a bit lower than I was, because the sound was so remote, so muffled, but maybe I was wrong. What would I say to him now?

But it was he who spoke: "You came in time. I knew you would make it. For me not to hear your voice again was unthinkable."

"Why don't you come out, Bern? Come back out of there, please."

The cold took my breath away. The air inside the cave was dense, laborious to inhale.

Oh, Teresa, how I wish I could. But I'm afraid it's too late for that. I'm no longer able to. I must have fractured something when I fell down here. The tibia, I think. And a rib maybe, although the pain in my side comes and goes. I haven't felt it for a few hours now."

"Someone will come to get you. Someone can come down and get you."

Jónas was somewhere in the dark. He had turned the lamp on his helmet off, perhaps to allow us an illusion of intimacy.

Bern seemed not to have heard me.

"There's a high, smooth wall on this side, like a plate of silver. A fine veil of water is trickling over it. It's almost like a mirror, I can make out the shape of my head when I project the light in a certain way, though the battery won't last much longer. How I wish you could see this wonder, Teresa. You know what I'll do? I'll pretend that the face I glimpse is yours and not mine. Would you do something for me?"

"Of course," I murmured, but he couldn't hear me like that, so I shouted it.

The strangest goodbye in the history of the world, forced to yell what we would otherwise have whispered.

"Look around you. Choose a shape, a rock that looks like a face, that looks like me."

I aimed the lamp's beam along the wall of the cave, frantically, not recognizing anything but edges and projections and bulges in that chilling place.

Bern remained quiet, giving me time, then he said: "Did you find one?"

"Yes," I lied.

"Good, so now you can look at me. Can you hear the sound of the drops? You'll hear it if we remain silent for a moment. They're like notes, the notes of a xylophone being faintly struck. But you have to turn off the light, so the mind won't be distracted by what the eye sees. Sight always seizes all of our attention, Teresa. Shh, listen now."

I did as he said. I fumbled with the lamp's switch until it went out. The cave plunged into total darkness, the most absolute darkness I had ever experienced.

After a few moments I heard the plink-plopping of the drops. Some produced sharp clicks, like wooden sticks, and others emitted notes at regular intervals. New ones were constantly being added, as if my brain were slowly becoming accustomed to capturing them, as if my ears were picking them out of the silence. In the end the sound became full to bursting, a concert of hundreds of tiny instruments, and I felt as if I could see again, but with a faculty that I had never used before.

"Did you hear it?" Bern asked, his voice now a roar compared to the drip drip drip. "Such a thing can only have been created by God."

"Do you believe in God again, Bern?"

"With all my heart and soul. I never really stopped believing. Although now it's something different. It's throughout my entire body, inside and out. I no longer have to make any effort. Do you know the saying, Teresa? 'I fled from your hand to your hand.' Do you know it?"

"No. I don't know it," I said brokenheartedly.

"It was one of Cesare's favorites, when we disappointed him. Sometimes we did it on purpose. He'd pretend not to notice, he knew we'd come looking for him again. And when we did, he would whisper those words in our ears: 'I fled from your hand to your hand.'"

He took long pauses between one word and another, as if he found it hard to breathe.

"Tell me about the masseria, Teresa. Please. I miss it more than you could ever imagine. There's not much that I regret in this cave, apart from not seeing you. And the masseria. Tell me how it was when you left."

"The figs were ripe."

"The figs. And did you pick them?"

"As many as I could."

d the holly oak? Were you able to make it recover?"

"Yes."

"That's good to hear. I was very worried. What else? Tell me something more."

But tears kept me from saying more, my throat was strangled.

"Even the pomegranate produced a lot of fruit," I shouted toward the fissure.

"The pomegranate," he repeated. "We'll have to wait and see, though, at least until November. You know how that tree is. It always promises glorious fruits and a week before ripening they split open. Cesare used to say that. He said there was something wrong with the roots. Maybe it's the proximity to the pepper tree, but I'm not sure that's it. You have to cover it when the first cold spell arrives."

"I will."

"Do you know what my favorite moment was? Our walks. Around sunset, when we had finished our chores. You always dawdled a little while I waited on the bench. Then we walked together on the dirt track. Past the iron bar, we usually turned right, though not always, sometimes we went left. But we never hesitated. We always knew which way to go, as if we'd made up our minds beforehand. And if the figs were ripe, we'd pick them even from the trees that didn't belong to us. Because in reality it all belonged to us. Right, Teresa?"

"Yes, Bern."

"It all belonged to us. The trees and the stone walls. The heavens. Even the heavens belonged to us, Teresa."

"Yes, Bern."

All I could manage to repeat was, "Yes, Bern," because my mind was racing ahead, to the moment when I would no longer be able to hear him.

From the darkness where he was watching over me, Jónas said it was time to go. But how could you decide when to end a time like that? How could we break off that conversation and leave Bern alone? I knew I couldn't hold up much longer, though, my feet were frozen stiff inside my boots. I couldn't move my fingers anymore.

"I need to ask you something, Bern. About Nicola."

He was silent for a while, then he answered very evenly: "You have to speak louder. I can't hear you that way."

Had he really not heard me, or did he just want to make me repeat it? Maybe he knew that my courage was already petering out; he still knew me better than anyone else.

But I was able to say it a second time, to shout it so he couldn't pretend not to hear me, and the re-echoing in the cave slammed my doubt against every boulder and hurled it back at me, multiplied. "I need to know about Nicola. Was it you, Bern?"

I pictured his close-set eyes open in the darkness, his expression. I had no need to find a rock that looked like him, he was imprinted inside me.

"I'd like to tell you a lie and swear it wasn't me. But there will be no more lies, I promised myself."

"But why did you do it, Bern, why?"

"Something compelled my foot. Something powerful. Nicola's head was on the rock and that force raised my foot and drove it down. The Lord stopped the hand of Abraham, but out there, in the olive grove, He did not stop me. There was no God at that moment, there was His opposite there with me, and he rammed my foot into Nicola's head. I'd like to tell you that none of this is true, Teresa. It's what I wish more than anything in the world."

"He was your brother. I don't understand."

"He . . . The two of you . . ."

"But that's not true, Bern! It's not true! You were the only one."

"And he'd said those words."

"What words?"

But he was silent again now.

"What words, Bern?"

"He was the one who gave her the leaves. He picked the oleander leaves and put them in her hands. He did it to protect himself."

"Which leaves? What are you talking about, Bern?"

"Sometimes we lose who we are, Teresa."

Jónas's light went on at the back of the cave, he came toward me. "We have to leave now," he said.

"No."

"We have to leave!"

Somehow he dragged me out of there. Getting out proved to be more difficult. I had no strength left, the result of the cold and my grief. I tried putting my foot in the toehold that Jónas showed me, but the boot lost its purchase; I no longer had any sensation. I slid back until he stopped me with his hands. He said we had to hurry, I was in danger of hypothermia.

Bern's voice filled the chamber once again:

"Will you come back?"

I promised him I would. We made our way through the cave, among the fragile stalagmites of ice, crawling up the slope on our bellies, stooped over our knees in the tunnel, and this time Jónas did not let go of my jacket sleeve the whole time, as if he were afraid of losing me.

After that there's a gap in my memory, until the moment when I found myself lying on one of the huge rocks in the lava field, under two layers of blankets, the sky once again above me, and that strange night that was too light. Giuliana stood over me, studying me. She said I had lost consciousness climbing up the metal ladder, and had very nearly tumbled down.

When I managed to sit up, they made me drink small sips of coffee. Half an hour had passed, maybe less.

"He'll die," I said.

Giuliana looked away. She poured more coffee into the thermos cap. "Drink some more."

"How has he been able to survive all these days?"

"He has good equipment. Food. Water. He was prepared to stay in there a week. And his endurance is incredible."

"But why don't they get him out?"

"Nobody is able to get in there. And even if someone managed to, they wouldn't be able to help him."

"They can drill through the rock. Make an opening."

Her eyes blazed. "The cave is a protected place!"

"But Bern is in there!"

Giuliana put a hand on my cheek, a cold, dry hand. "You'll never understand, will you?"

WE RETURNED to the lake in that slowly declining twilight. The two other young men came with us. The way back seemed shorter to me than the trip out had been.

There was a room for me in the apartment where the guides lived. It was as spartan as a hospital, the quilt folded on top of the bed. The dinner hour had passed; Giuliana said we wouldn't find anything open, but there were snack machines on the ground floor if I was hungry.

I stood under the shower for a long time to drive out the cold that seemed to have penetrated my bones. When I stepped out, the whole room was full of white steam. I didn't even have the energy to take clean underwear out of my bag; I wrapped myself in the quilt, naked, and fell asleep.

That night I dreamed about the masseria. I couldn't get in the house because the door was locked, but I knew Bern was in our room, lying on the bed. I called him from the yard, but he didn't answer. At one point a pebble was thrown out of the open window. I picked it up from the ground and threw it back. Maybe Bern had chosen that way to communicate with me. Then more pebbles began shooting out from the window, fistfuls of them. In the end they rained down from the sky as well, a black, hammering hail that in seconds buried the house and covered the countryside, leaving me in the middle of an endless desert.

IN THE MORNING we returned to Lofthellir. Only one of the young men who had been watching the cave entrance the day before came with us. He sat in front and talked to Jónas the whole time, shreds of conversation in that guttural, primitive, hateful language. Sometimes they laughed, but they seemed to quickly restrain themselves, as if aware that it seemed insensitive toward me.

At breakfast Giuliana had approached my table. Before setting down the skimpy dish she'd prepared for herself, she'd asked me if I would prefer to be alone. I told her to sit down, though not too politely. We exchanged a few words about nothing, about how unthinkable it was for Italians to eat smoked herring at that hour, even after one had lived there for months.

In the jeep, however, we managed to talk again. I asked her why Iceland, why that cave, why the inaccessible fissure in that cave.

"It's because of something that Carlos said."

"Who is Carlos?"

Giuliana tugged her sleeves down until her fingers disappeared.

"A guy from Barcelona. After Freiberg we went there. We had made contact with a group."

"What kind of group?"

"A little of everything. Independents, black bloc activists waiting for a pretext to riot. We drove there in a rented car. We thought we were being pursued, so we did it nonstop. By some miracle we didn't encounter any roadblocks. But we didn't stay long, I didn't like the situation there, and Bern in particular worried me."

She stretched her legs under the seat. She stared at them for a few seconds.

"He refused to leave the apartment. 'It's all so sick out there,' he said. 'Don't you see it? Don't you see how by now we've ruined everything?' They were things we'd discussed a million times, but now he meant something different, something I couldn't fully grasp. One day he began talking about how he had slept in a tree with his brothers. He had persuaded them to stay outside to see the shooting stars. Staring at the dark sky, he'd felt he was part of something that surpassed him. It was a very detailed account. At that moment I felt the frightening immensity of the love he had inside. It wasn't just about the trees, it was about everything and everyone, and it didn't let him breathe, it was suffocating him. Does that seem crazy to you?"

It didn't seem crazy to me. It was the most accurate description of Bern that I had ever heard. So Giuliana genuinely loved him. But that thought no longer upset me. Now, I simply accepted it.

"Anyway, one of the leaders of this Catalan group came to see us and that was the breaking point. This man, Carlos, had worked on Greenpeace ships in the Arctic. They talked for a long time. Bern was captivated. It was Carlos who first mentioned the Anthropocene to him."

"The Anthropocene?"

"The geological era we live in, in which everything on the planet, every place, every ecosystem, has been altered by man's presence. A concept that I had already heard talked about elsewhere, though Bern hadn't, and for him it was like a revelation. In the days that followed he spoke of nothing else. The desire to find at least one exception began to grow in him. Something that had not yet been seen, ruined. Something pure."

"Is that why you came here?"

Giuliana gave me a patronizing look.

"Iceland is the exact opposite of purity. The Vikings cut down all the trees on the island centuries ago. From a certain standpoint, Iceland is the ultimate realization of the Anthropocene, even though people come here to look for uncontaminated spaces. That's why Carlos mentioned it. He said 'Iceland,' but he could equally have said 'Amazon rainforest.' Bern took it as a dictum. So we came here to look for an exception. Our money ran out quickly; in less than two weeks we were broke. For a few months we worked on a farm by a fjord. A terribly isolated place."

Jealousy flared up again for a moment: Bern and Giuliana in one of the painted sheet-metal houses, shrouded in fog, warm inside and frigid outside. The sex between them. I suppressed that image as best I could.

"After the winter we moved to the lake. We met Jónas and the others there. They needed more staff for the high season, people willing to do a little of everything. Sometimes the tours they organized were dangerous. But Bern still had his plan in mind. Together with Jónas we visited the island's most remote spots, and they were never good enough. The mere fact of being able to reach them was proof of it. Until we discovered the cave."

"But you can get into the cave too, there's even a metal gate."

"Only as far as you went. No one has ever been in the next chamber. Its existence was known, but access to it was too dangerous and difficult."

"So Bern decided to be the first."

"And probably also the last, considering the outcome."

"Why didn't anyone stop him?"

Giuliana looked quickly at me, then away again.

"Every one of those guys would love to do what he did. They wanted to find out what was in there, and that way they would at least have been part of the discovery. Having studied the air currents within Lofthellir, they're convinced that there is a way out. At some location in the lava field."

"So could Bern find a way out of there?"

"If he hadn't fractured a leg, maybe. Now it's out of the question." We rode in silence for a while. We were on the worst stretch of the road, the jeep lurching on the shock absorbers, but this time the violence of the jolting did not surprise me.

Maybe to dispel the agonizing foreboding that had come over us, Giuliana said: "The tourists have a lot of fun on this road. Some start screaming as if they were on a ride at the amusement park. Bern liked it, too. He was enthusiastic about everything he saw on the island. When he entered the cave for the last time, he was smiling. Even though he knew it could end badly, even though by then he was just a bundle of nerves and determination. I'd never seen him so content. Maybe only at your wedding."

To this day I don't know if Giuliana said that to make me happy, but at that moment I chose to believe her. "A bundle of nerves?"

"He'd lost more than forty pounds. A child couldn't have fit through there, let alone an adult like him. But he was sure he'd make it, and he was right. For months he'd studied the sequence of movements, the necessary contortions. We took measurements of the cleft, of every projection and irregularity as far as we could see with the flashlight, and he built a plaster cast exactly the same. He kept it in the yard. It's still there. It weighs a ton. From the room I would see him practicing."

"From your room?" I interrupted her. I couldn't help myself.

"From our room, yes," she said tiredly. "It was as if he were practicing a choreography. He wrote everything down in a notebook. And when he wasn't training, he would sit still, cross-legged on the ground, as if he were meditating or praying, waiting for even the last molecules of fat in his body to melt away. Fasting was no trouble at all for him. Once he told me that his uncle had fasted for one month straight when he was young, and therefore he could easily get by with a cup of broth and some fruit every day. It had become impossible to make him eat anything else. He saw manipulation in any type of food."

"He's always been obsessed with those things," I said. "Since he met Danco, at least."

And you, I wanted to add, but I didn't.

"Not to that extent," Giuliana countered. "By then he refused to eat tomatoes because they hadn't existed here in Iceland before men introduced them. Unfortunately, nothing edible existed in Iceland before being introduced by man. So he drank broths made from a local herb. If I was the one who cooked them, I would add some meat on the sly. I'm sure he was aware of it, but he pretended not to notice. He was extraordinarily acquiescent. You felt as if you could hurt him, crush him with a wrong word or two. And not just because he was so thin. Still, once he was ready to go in there, after all the training, after he and Jónas had altered his clothes to adhere as tightly as possible to his body to keep him warm, and after we smeared them with fish oil to help him slide between the rocks, he was happy."

THAT DAY TOO Giuliana waited outside. Maybe she had already said her goodbyes to Bern before I got there; I wouldn't ask her, not even later on. Jónas accompanied me again. I was more assured, and it took us half the time to reach the end of the cave. I went to sit on the flat rock again, inside that absurd, echoing confessional, and called to Bern.

He answered only after the third or fourth time, when my heart, fearful, was already pounding out of control. His voice seemed fainter, farther away, as if during the night he had slid a few more yards down the frozen incline, where I imagined him shrouded in darkness.

He did not say my name, not immediately, the first thing he said was: "It's so cold."

I asked him if he had tried to move his leg, to get up, but he didn't answer, as if something much more important were weighing on him. He said, "This wasn't the journey I wanted, Teresa. The journey I wanted was with you."

But it wasn't him talking. Bern was no longer there. A specter spoke in his place, an echo of his voice that had remained trapped in that cavity of ice and stone.

For a few seconds there was only the silvery plink plink of the droplets. Finally he cried, "Forgive me!" and those were the last words he uttered, the last sound able to rise to the top of the rock incline, pass through the fissure, and come down to me. As if he had lasted all night and all morning just to emit that one sound.

After that I called and called to him, I don't know for how long, until a light appeared in front of me, aimed right in my eyes, and arms wrapped around my shoulders, and somehow, maybe dragging me, Jónas took me out of there.

GIULIANA HAD MANAGED to get the tourist visits canceled for that morning, but it was peak summer season, and by afternoon everything was to resume as usual. Toward three o'clock a party of about ten people arrived. I watched them proceed through the lava field single-file, each carrying a helmet and boots in their arms. Did they know that there was a man trapped down there? A man who was dying?

Giuliana stood beside me while one of the guides repeated the explanation about how to conduct themselves in the cave, the same explanation I had heard the day before. I had the feeling she was keeping an eye on me, in case I were to do something rash. But all I did was approach the guide when he finished speaking and ask him if I could join the group. Jónas stepped in, treating me kindly but firmly. His staff member, the other young man, would call out to Bern. If Bern answered, then he would take me inside again.

An hour went by, it seemed much longer. I dug a furrow in the soil with the tip of a twig, covered the groove back up, then dug again, deeper. When the guide reappeared at the top of the metal ladder, it wasn't me he addressed but Jónas again. He shook his head, and I understood that Bern hadn't answered.

We went back to the jeep. I sat in back. Throughout the trip I harbored a dull rage toward the tourists, for their cheeriness. Giuliana sat beside me, but her presence gave me no comfort.

BY THEN there was no longer any reason for me to stay in Iceland, but all the same I changed my return ticket once and then a second time. All in all, I stayed in the apartment with a view of the placid surface of Lake Mývatn for two weeks. I called my father and asked him to go to the masseria to take care of the vegetable garden and all the rest. I couldn't tell him where I was, nor what had happened, but he realized that it had to do with Bern from the way I started crying on the phone, unable to stop. He would leave that same day, he promised. I said I would explain what he had to do once he got there.

I did not return to the cave. Each morning I dressed as if I were going there, I showed up at the jeep's departure point, but when the tourists started to gather, young couples with a passion for harsh climates, amateur speleologists, overweight women who would likely not even be able to enter, my nerve failed. I felt like an intruder. Then I'd approach Jónas or the guide assigned to the shift and remind him to call to Bern once inside. Eventually there was no need to tell them anymore, they reassured me with a nod, patiently. I imagine they'd stopped calling to him early on, but I clung to the idea that it wasn't so: there wasn't much left for me to hold on to other than that perseverance.

It still wasn't clear to me what Jónas knew about the reason Bern and Giuliana had ended up there, but when the time came he did not insist on reporting Bern's death to the authorities, as though he sensed that the man who had ventured into that forbidden part of the cave existed and did not exist for the rest of the world. As though he understood that nobody, except me, would come to claim him.

To endure, I took long walks around the lake, in one direction in the morning, the opposite way in the afternoon. Mostly I went alone, but sometimes Giuliana came with me. She had at least partially shed the reserve of our first hours together. I leaned over the water to see if there were any fish, but I never saw anything, only algae swaying below the surface.

THE NIGHT before I left, I was awakened by someone knocking at the door of my room. I stayed in bed, uncertain, wondering if it was a dream, but then there was another series of knocks. I got up and opened the latch. Giuliana was dressed in a jacket and boots.

"Put something on. Come outside, quick."

Before I had time to ask her why, she was hurrying down the carpeted stairs. I threw on some jeans and a fleece I had bought to get me through all that time there.

The guys were on the lawn. Jónas pointed to something high above. The sky was arrayed with shafts of brilliant green light.

"You never see it at this time of year. It's a miracle."

They all had phones in their hands and were looking for the best angle to photograph the sight; they were excited, though undoubtedly I was the only one witnessing that phenomenon for the first time. The green rays seemed to radiate from a precise point on the horizon and from there spread out through the air like smoke. I did not ask if that was the direction of the cave. But when Giuliana said, "It's for you," I knew that it was really so.

One at a time they tired of looking and went back into the house. Finally, Jónas and Giuliana also left. The lights in the sky persisted. If they changed, their movement was so slow as to be imperceptible. Back in the room, I raised the plastic roll-up shade so I could keep watching them. In the morning, when I woke up, they had vanished.

GIULIANA AND I shared a cigarette in front of the airport. I didn't want one, but I wanted to prolong that moment.

"Will you stay here?" I asked her.

She looked around at the landscape, as if deciding there and then.

"For now I can't imagine anyplace else. And you? Will you go back to the masseria?"

"For now I can't imagine anyplace else."

She smiled at me. She crushed out the lit part of the cigarette and put the butt, the filter that would take years to decompose, in her pocket. Everything came to an end, and sooner or later it would happen, even to the grief we had in common.

"Maybe you'll see me show up there one day," she said.

We brushed cheeks, shyly, then I went into the airport. When I turned around, she wasn't there anymore.

8.

Every morning, at the masseria, my father would say: "You don't need me here anymore, I should go back to your mother." But another day went by and he was still there: he had to help me pick the tomatoes, the door hinges needed to be repaired, or else he was inspired to create a chair with what he found lying around. I'd told him what had happened in Iceland, confusedly, surprised at how bizarre my account sounded, to the point that I began to doubt my own words. But he'd listened to all I had to say, and in the end he held me tightly for a long time, while I cried pressed against him, as I don't remember ever having done.

As the days grew shorter, darkness forced us to stop our work and we cooked together. After dinner we went to bed early. Though my anguish waited in the bedroom, I knew that my father was nearby, just down the hall. I could hear his snoring through the partly closed doors, the same sound I once found objectionable and that now kept me safe. Then I thought about Bern's words in the darkness of Lofthellir: "'I fled from your hand to your hand.'" It had happened to us as well, to my father and me.

When he actually did leave, I was ready to be alone. As I accompanied him to Brindisi he said: "You should inform his parents."

"I'm not sure about that."

"They're his parents," he repeated, as if that were enough to counter any objection.

Several more weeks went by. I didn't get many visitors, except those related to work: the vegetable orders on Monday and Thursday, routine maintenance appointments, the helper who came every other afternoon. The tail end of summer was mild, and autumn was reluctant to start; the eggplants seemed as if they would never stop producing, by then they were as tall as saplings. I was outdoors almost all day, always busy; I didn't mind it. When working, I was able not to think as much, and mostly only practical thoughts. Still, I would sometimes linger mesmerized in the middle of the food forest, staring at nothing in particular. Sooner or later, certain questions would become nagging: What happens now? From where do I pick up the thread? I was thirty-two years old, and that meant an ocean of time to fill. Would I remain on that land forever?

WHEN I SAW the car appear on the dirt track I was stacking wood against the tool shed. A small compact that I didn't recognize, its front end bashed in by a collision. I approached, slipping off my gloves, and when the car stopped I recognized Cesare. He nodded to me. Beside him sat his sister, who did not greet me until we stood facing each other; then she offered me her small, delicate hand, just as I remembered it from the night of the wedding.

"Would you like to come inside?" I asked. "I think it's going to rain."

Cesare opened his mouth wide and inhaled as much air as he could take in. He seemed to savor it, relish it. The smell of the masseria: I knew exactly what he was looking for.

"I'd like you to take me for a little walk around first," he said, beaming. "Yes, I'd like to see everything, if you don't mind."

So I led him through the property that at one time had been his, and explained all the changes to him, the way Bern and Danco had once explained them to me: the water canalization and filtering system, the wall of twigs and straw on which the aromatic herbs grew. Every piece of information seemed to wholly intrigue him. He listened to me with his hands clasped behind his back, then exclaimed: "Magnificent!"

Marina trailed behind us, gazing around blankly, and when he asked her what she thought, she hedged.

"You have resurrected this place," Cesare said at last, with the solemnity that was his and that would have been ridiculous in anyone else.

We sat under the pergola. He observed the tablecloth with the world map with a blend of amazement and perplexity, and maybe even nostalgia, then he turned that same gaze on me.

"I've never found another one that would do," I said. "Though maybe it's high time I did."

I brought out a carafe of water, an open bottle of wine, and some toasted almonds.

"I received your note," Cesare said. "We received it. Marina is very grateful to you for having informed us. Isn't that so?" he asked his sister, affectionately touching her arm; she agreed, reticent as before. "That boy was capable of extraordinary exploits," Cesare went on. "But the discovery of the cave astounded me."

"I didn't call you after Nicola's funeral. I'm sorry I didn't."

"Genuine sorrow is worth more than a thousand gestures, Teresa. More than all the phone calls in the world. And I knew you were grieving, I felt your sorrow with me."

They hadn't touched either the wine or the water. I should have filled the glasses, but I was somewhat dazed. "And Floriana?" I asked.

"Oh, my Floriana. Grief has poisoned her heart. I wish I knew the antidote to cure it, but I don't. Patience maybe. Time? I can't imagine that we will remain separated for much longer, you see. And perhaps the Lord will grant the prayer of an aging man."

He smiled. What he'd said was true: the last few years had taken a toll on his face, etching lines on his forehead and around his mouth; his kind eyes were somewhat sunken, his hairline had receded, and his hair was now medium length, not as if he meant to let it grow long as it once was, but as if there was no longer anyone to remind him that it needed a trim from time to time.

"And you, how are you coping?" he asked.

I was unprepared for such a straightforward question. "There's always a lot of work," I said.

Nodding to himself, Cesare seemed to be considering whether the answer satisfied him or not.

"When are you going to pick the olives?"

"I'm thinking of starting in November. But if the heavy rains come, it's possible I'll have to begin sooner. Water in September isn't good for olives," I noted, immediately ashamed I'd said it, of having been so arrogant. "But you know that better than I do," I added.

"There's a popular saying for that"—he squinted, concentrating—"but I don't think I remember it anymore."

All that hollow hypocrisy, that conversation skirting the edge of the precipices that affected us, upset me, especially with him. But we kept on. Cesare asked if I planned to press only the olives picked from the trees or those that dropped to the ground as well. I told him I would sell those on the ground to a local oil press.

"Then you will certainly get an exceptional quality," he said, and after that we drifted into an awkward silence. I saw him catch his sister's eye, as if asking her permission, and I saw her tighten her lips nervously.

"Marina and I," he resumed in a more serious tone, "came to ask a favor of you. We understand that the circumstances of Bern's death do not allow us to have his body here, physically, and be able to bury him. But you know how important it is for us. Burial is the only way the soul can be set free to look for a new abode. Remember when we buried the frogs out here? The first time you came to visit us at the masseria?"

"Yes."

"Well, Marina and I are certain that Bern would like his symbolic burial, all we are able to do, to take place here. Don't you agree?"

"We don't know if he's dead."

"From what you wrote to me, from the way you described it in your note, I seemed to think he was."

"It's not possible, I'm sorry," I repeated, more determined this time, but looking at Marina instead of him.

"It's a great torment for souls to remain abandoned in a body that no longer functions," Cesare persisted. "They are prisoners."

"I understand." I hesitated, then I said it: "But those are only your ideas."

And yet his words had evoked with heart-wrenching clarity the image of Bern lying in the cave's dark chamber, his fractured leg forming an unnatural angle on the ice, the skin of his face rigid, his wide-open eyes the same color as the air and the rock. Bern immune to any evolution or deterioration, for all eternity.

"Would you excuse us a moment, Marina?" Cesare said, standing up. "Teresa, come with me, please."

"Where?"

"You didn't take me to see the holly oak, after all this time. Let's go and sit there for a bit."

I followed him. Watching him walk stiffly in front of me, I realized that the problem with his hip had worsened, that his gait was irregular. Each time he placed his left foot, it was as if he were falling over it.

We sat down on the bench. Cesare reached out a hand to pick off a leaf, studied its outline, then, frowning, looked at the trunk.

"I'm treating it. The gardener says it's already better."

"Thank heavens. What an inestimable loss it would be."

He waved the leaf by its stem, one way and then the other.

"Bern and the people he was with during that last period," he said, "had a special veneration for trees, didn't they?"

I nodded.

"I read something about it in the newspapers, but I don't think I fully understood it. I sense that they weren't wrong, but I wish Bern had talked to me about it. Maybe we would have arrived at a different solution. We communicated very well together, the two of us. He was always immensely gifted in matters of faith, intuitive, but he tended to be somewhat reckless. Trees can inspire a sense of sacredness in us, I don't deny it, but they are not endowed with the same soul we have. Yet how magnificent they are. Majestic. Look up there."

I did it, even though I was already familiar with that angle, I knew it in every season of the year.

"You're keeping something from me, Teresa."

"No," I blurted, maybe too hastily.

We were silent for a very long time. I kept looking toward the house. Cesare's torso swayed faintly, back and forth, the leaf still in his hands. I had the feeling he was smiling, but I didn't dare check to see.

And finally what he'd been waiting for from the beginning happened. The confession spilled out at a moment when my mind was perfectly blank, defenseless: "It was him, he killed him."

It was the first time I'd uttered those words. I hadn't even been able to speak them to my father. They ignited the afternoon air.

Cesare laid a hand on mine. "My poor Teresa, what a weight you've had to carry. I know how much you both loved each other."

He took a few strangled breaths. Then he said, "I truly think our Bern would want a burial here."

I looked at him then, I was stunned. "Did you hear what I said?"

"I heard it."

"Then why do you care about his burial? It makes no sense."

He bent his head back to look up again. He closed his eyes, and when he opened them, they seemed filled with gratitude. I had a sharp recollection of how he had been when he was young, all the kind wisdom that emanated from his body.

"Because he's Bern. My son."

"But he killed Nicola! He was your son! How can you forgive him?"

"Think about it, Teresa. What would everything I taught you all mean, if I could not forgive Bern now?" He searched for words for a moment, then he recited: "'*If my brother sins against me, how many times must I forgive him? As many as seven times?' And Jesus answered: 'I say to you, not seven times but as many as seventy times seven.*' Seventy times seven. I haven't even started, you see? I really hope you will help me."

I struggled to regain my composure. "The people there knew that the cave had a way out. They were sure of it. He could still be alive."

Cesare looked at me intently. "Your hope is moving, and the Lord will certainly reward you. All I ask is that you think about it. If nothing were to change and you should feel it was the right time."

"Why don't the two of you do it? If it's so important. You don't need me."

"I'm afraid it wouldn't be the same. You are his wife. It is you, more than anyone, that he would like to have present."

"Shall we go back to Marina?"

I didn't wait for his response. I preceded him to the pergola.

"Are we ready to go?" Cesare asked his sister.

She stood up. She gave me her hand as before, but this time she leaned toward me and kissed my cheek.

"I would have liked to know you better," she murmured.

I picked up the bowl with the almonds, as if bringing it inside were urgent, then I stood there stupidly and put it back on the table. Marina took one and chewed it. "They're good," she said.

I walked them to the car. Cesare fastened his seat belt before starting the engine. "Goodbye, Teresa," he called through the lowered window.

But now I was the one who wasn't ready to let him leave.

"Bern talked about some oleander leaves."

He frowned. "I wouldn't know."

"Maybe he was just confused, but it seemed important. Something Nicola also knew about. Something serious."

Then his eyes slid toward the olive grove. More precisely, though I wasn't yet able to make the connection, toward the reed bed, rustling unseen behind the trees.

"He was talking about that girl, probably. Violalibera."

The same old uneasiness and fear exploded in my stomach.

"Violalibera?" I repeated slowly.

"She was a pitiable creature. And the boys were still so young. Bern was never the same after that. I was sure he had told you about it."

"Yes, of course . . ."

After which they left. If something like revelation really exists in the world, for me it occurred at that instant, as Cesare's car disappeared down the dirt track, the intensity of his presence still filling the air, and as I heard that lost name, Violalibera, a name that had reappeared after years, popping out of the soil like a capricious weed. The name, it was suddenly clear to me, that encompassed the inextricable tangle of our lives.

THAT SAME NIGHT I went to see Tommaso. The thought that he had a right to know what had happened to Bern nagged me, but I couldn't make up my mind to face him. Now I couldn't put it off any longer: if there was anyone who could clarify once and for all what had happened with Violalibera, it was him.

It had started to rain after all, catching drivers off guard, and creating a bit of chaos, so I found myself stuck at the entrance to Taranto. The lagoon on my left was a flat black expanse. I turned on the radio, but the music made me nervous, the voices and commercials made me nervous, so I turned it off and let the din of the rain against the roof of the car wrap around me.

I parked the car at an angle, partially blocking someone's gate, and switched on the emergency lights. I wouldn't be there long, only as long as necessary. The names on the buzzer panel were mostly foreign: Slavic, Arab, Chinese, grouped as many as six or seven on the same label. Conspicuous in the center was a piece of yellow paper, badly torn and affixed with scotch tape, the initials T.F. written on it. I rang and Tommaso buzzed me in immediately, without asking who it was.

I didn't know what floor, so I went up on foot. When I got to the fourth floor, the timed lights suddenly went off. The door to my right was partially open and a reddish glow filtered out. Voices could be heard from inside, and I approached them. Four men were seated around a table covered with a green cloth, playing cards. The cigarette smoke had formed a foggy haze. Tommaso appeared in front of me. He had bills in his hand and looked taken aback.

"What are you doing here?"

"You buzzed me in."

Laughter broke out in the apartment. One of the men said something, and the other voices rose over his. Tommaso slipped out. In the last fraction of a second when he wasn't blocking my view, I glimpsed a woman in shorts, long bare legs, blond hair loose down her back.

"You have to go!" Tommaso said.

"Who are they?"

"It's none of your business. People."

"I can see they're people."

"I'm working."

"Is this what you do?"

"Do you mind telling me what you want from me?"

He grabbed my shoulder, but the contact troubled both of us, and he immediately withdrew his hand.

"Violalibera," I said, letting the name produce a reaction on his face.

"I don't know what you're talking about."

He quickly swung open the door and dashed inside. I just had time to stop it with my hand before he slammed it in my face.

"Tell me what happened, Tommaso."

"Ask Bern what happened, if you're so interested. Now go."

"Bern is dead."

The words I had done my best to deny to Cesare a few hours earlier I spat out at Tommaso that night. Instantly, the life went out of his eyes. He bowed his head slightly.

"Never come here again," he said softly.

I took my hand off the door and he closed it. I heard one of the men ask where the pizzas were, then there was more laughter. Soon Tommaso would open up again, he'd question me to find out exactly what had happened, he'd beg me to come in. I just had to wait a little longer. I looked for the switch on the wall.

A few minutes more and the lights went out, I turned them back on. The elevator started moving and someone got off at an upper floor, there was fumbling with the keys. What was I thinking, asking him if that was the work he was doing? What gave me the right? What Tommaso did hadn't been my business for some time, and probably had never been. When the lights went out for the second time, I left.

AFTER THAT ENCOUNTER I fell into a kind of sickness. Now it seems natural to call it that, a sickness, but in those weeks everything about it seemed perfectly normal. I kept seeing Bern. Not clearly, not there in front of me in the flesh, but rather a heralding, as if each time I were about to see him in flesh and blood. It happened especially when I returned to the masseria in the car. There was a precise instant, just before turning onto the dirt track, when I was sure that I would find him waiting for me in the yard, sitting sideways on the swing-chair, or standing up, his back to me. His position changed, but the way he appeared in my mind was always just as detailed, the certainty that went with it just as intense. When I came out of the ground-floor bathroom. When I stood up straight after having been bent over for a long time in the greenhouse. When a window slammed. At each of those moments I knew without a shadow of a doubt that Bern would be there. I'd say to myself, There he is, not at all surprised. What surprised me, if anything, was not seeing him there shortly thereafter. But even that disappointment was slight, as if he were simply late, or somewhere else, close by at any rate.

I was not concerned about how lucid those presentiments were. Yet a sense of caution kept me from talking to others about them, from seeing others in general. When December came, I told my parents that I would not be back to Turin for Christmas. Maybe later on, I promised. I must have seemed myself, because they did not insist.

I decorated the holly oak with four garlands of lights and they were my only preparations. Despite my lack of interest, on Christmas Eve day I found myself battling with a sadness that seemed to lay siege to the masseria. Around seven I was lying on the couch; darkness had crept into the house since I'd lain down there, and I was considering the possibility of not moving until the following day, until Christmas had passed and everything was back to normal.

When the phone rang, I didn't rush to get up, but let it go on ringing for a while.

"It's me," a voice said, then added something barely comprehensible, as if the caller had abruptly moved the receiver away from his mouth.

"Tommaso?"

Silence.

"Tommaso, what's wrong? Why did you call me?"

I heard him take two deep breaths. "Oh. Teresa. I hope I'm not disturbing your dinner."

Was he giggling? The glow from the blinking lights around the holly oak intermittently revealed the objects in the room.

"You're not disturbing anything."

"I thought so."

"Did you call me to make fun of me?"

"No, no. I'm sorry. Absolutely not."

More deep breaths, then some coughing. He'd moved the receiver away from his mouth again.

"I'm waiting for Ada," he said after clearing his throat. "Christmas Eve is my turn this year. But I think I'm sick. Well, I wondered, see, if you could come and look after her for me."

So he needed me. After turning me away from his house, he needed my help. I let a few seconds go by.

"Well?" he pressed me.

As much as I wanted to be unfriendly, I couldn't do it. Was there really no one else he could call?

"I can come over," I said.

"Ada will be here in an hour."

"I don't think I can get there that soon."

"Well, do the best you can. It would be best if she didn't see me like this."

In the dark I looked for my shoes and jacket, then the car keys. As I did so, I knocked over a cup of pens on the desk, but I didn't even think about picking them up.

WHEN I GOT to the fourth floor, the door was partly open as it was last time, but there was silence. I entered warily.

"In here," Tommaso's voice called from another room. He was lying with his head slightly elevated; there were shadows under his eyes and his face was ashen.

He tried to raise his head, but the attempt made him grimace. I noticed the edge of a plastic basin under the bed and recognized the stench that filled the air.

"You're not sick. You're drunk."

"Uh-oh . . . Caught!"

He gave a lopsided smile. A dog was curled up on the empty side of the double bed. He looked at me with a resigned air.

"Why didn't you tell me on the phone?"

"I was afraid you wouldn't feel as sorry for me."

"I didn't come because I feel sorry for you."

"Oh, no? Why, then?"

"Because . . ."

But I couldn't finish the sentence. Because we're friends?

"What a model father, huh?" Tommaso said. "Christmas with hung-over Daddy. It's enough to call the social workers. Corinne can't wait."

He tried again to sit up, but he got so violently dizzy that I had to catch him so he wouldn't topple off the bed.

"Stay down!" I said, beginning to panic. "I don't know what you could have drunk to put you in this state."

"I broke all the rules of the responsible drinker," he said, and groaned, pressing a palm to his forehead, as if to stop something from spinning woozily. "Don't mix alcohols, don't switch to a lower proof, don't drink on an empty stomach. And above all, don't start before five in the afternoon."

"What time did you start?"

"At six, actually. But six yesterday."

Again, the same little giggle that I'd heard on the phone.

"I've never seen anyone so drunk."

He very cautiously removed his hand from his forehead, as if wanting to make sure that his brain stayed put when he moved it. "Then it really has been a long time since we've seen each other, Teresa."

He asked me to lock him in the room. He didn't trust himself to do it from the inside. I had to swear that I wouldn't open the door for any reason while Ada was there, even if she raised holy hell to get in.

"If she sees me like this she'll tell her mother, and if she tells her mother . . ."

"Right, I get it. Should I go down and get her when she arrives?"

"She comes up by herself, that way Corinne and I don't have to cross paths. Please don't say anything into the intercom. Just press the buzzer. If she hears a woman's voice . . ."

"But she'll tell her that there was a woman here."

Tommaso pounded his fists on the mattress. "You're right. Shit, what a mess! What a fucking mess!"

"Take it easy."

He reeked of alcohol, repulsive. His eyelids quivered. After bringing him a glass of water and locking him in his room, I did my best to make the living room presentable. Before that night, I'd thought that collections of empty bottles were an exaggeration seen in films, but now I discovered them in unexpected places throughout the house. I put them all out on the balcony. Medea, the dog, followed me seraphically. Tommaso had wanted her to stay with us, because her presence would reassure the child.

From the intercom's black-and-white monitor I saw Ada say hello, beaming, convinced she was talking to her father. Corinne stood a few steps back, only her legs could be seen. I pressed the buzzer without saying anything.

Ada was evidently allowed to go up the stairs alone, but not to use the elevator. I listened to her steps getting closer and closer. She'd started out running, but had now slowed up. On the landing I switched the lights back on every time they went out. At those moments she stopped, maybe surprised by that small miracle.

Would she even remotely remember me? Probably not. When she appeared on the landing, so pretty in a woolen beret with a pom-pom, her complexion fair though not as pale as her father's, I was sure of it. Her eyes revealed the uncertainty that she had mistaken the door, the floor, the building, the day, and she didn't know what to do. She was about to turn back, but I told her: "He's here, Ada. Don't worry."

She started when she heard her name.

"I'm Teresa, a friend of your dad. He's not feeling well tonight, he has a bit of a fever in fact, so I'm here."

She was still hesitant: beneath the beret, all those grave admonitions not to talk to strangers were whirling around in her head and, at the same time, there was no other choice but to trust me.

"We've already met," I said.

Ada slowly shook her head.

"But you were very small, about this big."

Something about that gesture encouraged her, because she finally let go of her grip on the handrail and took a step toward me. When we went inside, she checked to make sure it really was the apartment she knew. Then she ran to the door of Tommaso's room and tried in vain to open it.

"He's resting now. You'll see him later, I promise."

But Ada kept doggedly jiggling the handle. Fortunately, Medea came out to her from the kitchen and barked a couple of times, letting the child pet her and rub her cheek on her nose.

I took advantage of the moment. "Would you like to bake some cookies for Santa? We'll put them at the window with a cup of milk."

No answer, not even a polite glance on her part. I had been able to deal with whole classes of kids and now a single child, with her silence, was capable of making me die of mortification. She flopped onto the sofa, still dressed in her coat and beret. Just then Tommaso's snoring could be heard through the wall. I absolutely had to say something, to cover up that noise, so I went on blabbering about Santa Claus, about when he would come in through the window, I myself didn't even know what I was saying, but I spoke loudly and somehow managed to get through to her, because when I stopped she was different, more composed. She said, "I'm hungry."

Perfect, something to do, a way to move us out of there and into the kitchen. I persuaded her to take off her coat. I opened the fridge and the pantry. There were more empty bottles.

"Pasta with olive oil. That will be our Christmas menu. What do you think?"

Ada nodded and I thought I saw a trace of a smile. At the table she ate, staring at the small Christmas tree in the corner. Every now and then she slid a hand under the table to offer a crust of sliced bread to Medea.

A COUPLE OF HOURS LATER, when she'd fallen asleep on the couch, her jaw contracting rhythmically, I took the key out of my pocket and opened the bedroom door.

"Is she sleeping?" Tommaso asked me.

"Yes. I thought you were asleep too. Or that you were dead. I was a little worried."

"I'm awake. As for alive, I can't guarantee it. How did it go?"

"Good. We baked cookies and drew pictures."

"She's a sweet girl," he said. The drunken binge seemed to have drained him.

"You should drink some water. I'll bring you some."

I placed a full glass on the bedside table. I straightened out the sheet and bedspread, then I made him raise his torso for a moment so I could put the second pillow behind his head. Tommaso watched curiously as my hands moved around him. "I would never have imagined this," he said.

"Me neither, you can be sure of it."

When I thought he was comfortable, I looked down at him. "Viola-libera."

Tommaso closed his eyes. "Take pity on me."

"I can go and wake her up right now."

"You'd never do that."

So I called out his daughter's name, not as loud as I could have, but loud enough to really wake her up. Tommaso jumped.

"Stop it! Are you crazy?"

"Violalibera. I'm only repeating it once. Then I'm going to call Corinne."

His rage at me rose in him with all its old intensity. He clenched his fists on the bedspread.

"Okay, you win."

"I'm waiting," I said, afraid I'd lose my resolve any minute now.

"Get that chair," he said, pointing to one next to the closet, buried under his clothes.

"Will it take that long?"

"Get it. It makes my head ache to look up at you standing."

I went to the chair, grabbed the clothes in one armful, and dumped them on the floor. Then I moved the chair beside the bed. Tommaso's eyes were closed again.

The apartment was wrapped in silence, broken only by Medea's moist panting and Ada's slightly more restless breathing in the other room. For a while nothing happened. Tommaso opened his mouth once, but hesitated. Maybe that wasn't the point he wanted to start with. It would take a lot more time than I imagined.

"The institution," he said, "was a horrible place."

EPILOGUE

THE
DARK DAY

Many years ago, my grandmother had told me that we never fully know someone. I'd been in the pool up to my hips while she, lying on the lounge chair, squeezed the puckered skin of her knees, observing what her body had become.

"You never stop learning, Teresa. And sometimes it would be better not to start at all."

That afternoon I hadn't paid attention. I was eighteen and I was intolerant of any advice. My mother always reproached me for being impulsive and obstinate, a combination, she said, that would lead to no good. But my grandmother's words stayed with me somehow, and after the night at Tommaso's house, that long night of wakeful vigil and grievances, with him bedridden, I often found myself thinking about them.

"We never fully know someone . . . It would be better not to start at all."

The truth about people. That was what she was referring to, I think. Is there ever a point at which we can say we know it? The truth about

Bern, about Nicola and Cesare and Giuliana and Danco, the truth about Tommaso and again about Bern, especially about him, as always. Now that I've filled in the gaps of his story, of our story, can I say that I really know him? I'm sure my grandmother would say no, that any sensible person would say no: because the truth about people, about anyone, simply doesn't exist.

And yet, in spite of everything I'd learned about Bern from Tommaso and Giuliana, from all those who had the privilege of being with him when I wasn't there, my conviction remained the same as it had been before, my answer identical to the one I hadn't given my grandmother for fear of disrespecting her: I know him. I knew him. As no one other than I did.

Because everything there was to know about Bern I'd known just like that, having seen it in the first look he'd given me from across the doorway of the house, when he'd come to apologize for a ridiculous infraction. The truth about him was wholly revealed in his dark, close-set eyes, and I had seen it.

ON CHRISTMAS MORNING, when I woke up, Tommaso was not in the room and the door was closed. The sheets on his side of the bed were a tangle, the pillow folded in half. Maybe nausea had made him sit up again. A wintry light flooded the room with dust motes. All that remained of the turmoil that last night's story had triggered in me was a backwash of exhaustion.

I heard Tommaso's voice, then Ada's ringing tones, from outside the door. Something bouncing on the floor a few times. Then the intercom buzzed and they went out. Silence. So I got up and raised the shutter. The solidity of the objects I touched struck me as something new. I opened the window and the December air flowed in.

Four floors below, on the sidewalk, stood Corinne, wearing a cream-colored coat. The elegance suited her. Tommaso and Ada appeared in front of her. I watched them talking. Then Tommaso bent down to kiss his daughter. When he straightened up, he boldly leaned toward Corinne. Their cheeks brushed lightly, and finally she walked off, holding Ada by the hand.

When Tommaso came back in, I was making coffee.

"I didn't let you say hello," he said. "It seemed best that she not see you here in the morning, it would have been a bit complicated to explain."

"How are you feeling?"

"As if they decapitated me and then glued my head on backward."

In fact, he still looked terrible. He leaned against the kitchen counter.

"She talked about you. About the cookies you baked for Santa," he said.

"I managed pretty well, I think. But the cookies were awful. You don't even have any butter in the fridge, did you know that?"

We drank coffee. I knew it was my turn. I didn't talk for long, though, I was not as exhaustive as he had been with me. I told Tommaso little more than what I had written to Cesare in my note. I described the fissure in the cave that Bern had found a way to squeeze into, as if he wanted to inseminate the whole earth, but I told him nothing about what Bern and I had said to each other through the damp rock. And I didn't tell him about Germany or about Bern's father or about Giuliana either.

Tommaso's expression never changed, he did not cry and in the end he did not ask me any questions. I went to look for my handbag. Tommaso asked me what plans I had for Christmas dinner.

"No plans, no dinner. And you?"

"Same."

On the landing I thought that would likely be the last time I'd see him, my most faithful foe.

"Thanks for saving me yesterday," he said. "I suppose one offers to reciprocate in such cases, but I wouldn't know how."

I didn't feel like going home, so I indulged in taking a walk. I strolled through the old city, among the dilapidated buildings and abandoned courtyards. I reached the swivel bridge and continued on past it. Even in the center, the bars and shops were closed; on the streets there were only families on their way home from mass, some with bouquets of flowers and bags full of presents. I found myself at Corinne's house without having decided it beforehand. From a distance I glanced at the windows and thought I spotted someone behind the glass. I missed Corinne, I missed her voice, her sardonic smile. Maybe someday I'd go and see her. Heading back, if I retraced my steps slowly enough, I would be safely past dinnertime. Not that solitude scared me, but it seemed easier this way.

When I turned onto the dirt track, almost two hours later, I was ready for the usual prefiguration of Bern, but that day it didn't happen. Wherever his ghost had been lodging during the last few months, whether in the countryside around the masseria or just in my head, that Christmas morning it had gone, and would not reappear. In the house, everything was exactly as I had left it the night before. The pens that had fallen off the desk were scattered on the floor, and the caps had popped off some of them. I picked them up and put them back neatly in the pen caddy.

BUT I WAS WRONG about Tommaso, about the fact that I probably wouldn't see him again. It was I who called him a few months later. Spring had already begun, I'd bought a big flowering hydrangea that I planted against the bare wall of the house, where the roof would guarantee enough shade. The hydrangea demanded a shameful amount of water, but I had always wanted to have one, and maybe I was beginning to be tired of the yard's austere aridity. After all, it wouldn't harm anyone, it wouldn't worsen the condition of the soil, but it would help me every time I looked at its lush globes of white petals.

I phoned Tommaso and asked him if his offer of returning the Christmas favor was still good. He said yes, but warily, as if he expected a request that would be difficult for him.

"Would you go on a trip with me?"

"A trip to someplace far away?"

"Very far. But your expenses will be paid."

In February I'd returned to Sanfelice's office in Francavilla. I had not made an appointment, I simply showed up. Then I waited for a brief opening between two visits, as the new secretary, a bright, efficient girl, studied me. If I had followed the usual procedure, I wouldn't have had enough courage to see it through. Instead, there I was.

When he saw me, Sanfelice straightened up in his chair, his face alarmed, a hand already on the phone to call for help.

"He's not here," I said, "don't worry."

He drew his hand back from the receiver, still uncertain. "The last time he came here he scared the patients to death. And me too, to be honest. See the Rolodex? He grabbed it and started destroying everything."

He shook his head to dispel that image from his mind. Then, realizing that he had left me standing, he invited me to sit down. I told him that I wanted to try again.

"Does your husband agree?"

"I told you, he's not here."

Perhaps he wondered if he should inquire further, but he decided not to. I explained to him that among the forms that Bern and I had signed in Kiev was a consent to freeze the embryos. Maybe they were still there.

"Well, we can ascertain that right away."

He took out an agenda and entered the number. He spoke a while in English with Dr. Fedecko, nodding to me.

AND SO IN APRIL, four years after taking that route with Bern, I again crossed the bridge over the Dnieper, its waters sparkling on a day that was still cold, and almost unbearably luminous. Boats moved slowly in both directions, fanning out the surface of the water.

I noticed that Nastja was throwing hostile glances at Tommaso through the rearview mirror. Since leaving the airport she had spoken very little.

"I know what you're thinking," I said, "but he's just a friend. Bern couldn't come."

"Oh, I don't interfere in other people's business," she replied, acting offended, but I knew the clarification had appeased her.

"I'm here because the dark day has come," I said.

"What dark day?"

"You were the one who told me that once. That you have to put things aside for when a dark day comes. And now it has."

She smiled at me. "Then I'm glad I told you."

AFTER THE TRANSFER, Tommaso tiptoed into the room where they had left me to rest.

"I'm not sleeping," I said. "Come on in."

He wore blue nylon booties over his shoes and a paper gown tied in back. His diligence was touching.

"See those cupolas up there?" I said, pointing to them. "It's the Lavra. Bern loved it."

But Tommaso was studying me intently, clearly apprehensive. "Do you feel all right?"

"Sure."

"So now what happens?"

"Now we go home. Would you hand me my clothes, please? They should be in the closet."

I think it was at that moment that I decided. As Tommaso gently helped me slip my arms into the sleeves of my sweater, slightly embarrassed maybe at having to deal with my half-naked body, I decided that I would grant Cesare's wish.

But I waited for May to go by, then June as well, so that by the time I called him and the appointed day arrived, summer was already at its peak.

Cesare arrived wearing a purple stole around his neck. "Which spot did you choose?" he asked me.

"The mulberry."

We set out for the tree, where Bern and his brothers once had their treehouse. Cesare and I led the way, with Marina following closely behind us and Tommaso farther back. Ada skipped around him.

The incessant chirping of the cicadas accompanied us among the olive trees; the setting was identical to my first summers there, when for me Speziale existed solely in that season.

Cesare asked Marina to hold his stole while he dug.

"Show me what you brought," he said.

I turned to Tommaso. From the side pocket of his pants he took a yellowish book whose corners were all curled up.

"I found it," he said.

From his hands Cesare took the copy of *The Baron in the Trees* that had been Bern's when he was a boy. He leafed through the pages, still crouched on the ground. For a moment he focused on an underlined sentence.

Placing the book in the small grave, he recited a psalm and then a passage from the Gospel of John: "'My Father's house has many rooms; if that were not so, would I have told you that I am going there to prepare a place for you?'"

He asked if anyone wanted to add something. We were all silent, eyes fixed on the cover of the book. And, since no one spoke, he began singing. He had lost some of the pitch he once had and at times his voice seemed to crack, especially when he strained for the higher notes, with that somewhat nasal timbre I remembered so well. But the resolve with which the song resounded in the blistering air, that had not changed at all. I thought he would continue solo to the end, but Tommaso joined him in the second stanza. They sang the rest of the song together.

I think Ada sensed the solemnity of the moment. Looking up, she watched her father sing, as if that simple act showed her something unexpected and very important about him.

We covered the hole back up. Cesare sent us to collect stones and placed them over the spot where the book was, forming a small pyramid. Goodbye my love, I thought.

WHEN CESARE AND MARINA LEFT, Tommaso and I walked among the olive trees, while Ada chased after one of the feral cats.

"Will you come sometimes?" I asked him.

I was sure he saw people and events from the past everywhere he looked, just as I did. "Ada likes it here," he said. "She seems to have grown attached to the place already."

"I'll need help from here on out. For free," I added.

Tommaso smiled. "For free, of course."

We didn't make any promises, however. That was just fine.

I told him about the shafts of green light that had appeared over the lake after Bern died. I hadn't told him before, but for some reason I felt I owed it to him.

"The Northern Lights are very rare at that time of the year, that's what they told me."

"But you weren't surprised."

"No, actually not. Sometimes I feel like a lunatic. Look what we just did, we buried a book."

Tommaso twirled his finger in the air, making the cuckoo sign.

"Maybe it's nuts," he said. "And probably what you saw was only an atmospheric phenomenon with precise causes. Only it's terribly sad to think so."

"You know what Danco would start shouting now?"

"Obscurantists! Evil reactionaries!"

"Goddamn backward-looking dinosaurs!"

We laughed. Then Tommaso said: "I heard he returned to Rome."

"Yes, that's what I heard, too."

A magpie flew up from the ground and went to perch on a branch. For a moment our gazes met up there.

We played with Ada some more, then they too left. I sat down on the swing-chair. I had sudden moments of exhaustion, as if all the blood in my body were abruptly drawn to a single location. Sanfelice had told me it could happen, especially in the early months. I waited for those minutes to pass.

The sun had eased its hold, now the light was so hypnotic and perfect that I wanted it to stay that way forever. It was the time of day when you fell hopelessly in love with that place. I remembered the emotion that overcame Bern every time he gazed deferentially at the countryside at sunset. Would that emotion be handed down? Was it written in some

sequence of the genetic code, or would it vanish? I didn't know. But I hoped it would not be lost. All I could do, someday, was tell my daughter who her father was, try to explain to her what he'd revered and the mistakes he'd made doing so. Convey what, in his short life, he had loved of the earth and of the heavens, unremittingly, with all the abandon and impetus that are granted to a man.